高等教育安全科学与工程类系列教材

安全工程专业英语

第 2 版

主　编　司　鹄
参　编　张东明　谢　波　严　波
主　审　吴　超

机械工业出版社

本书选编了安全科学与工程学科涉及的科技英语文章，内容包含安全管理学、系统安全工程、安全人机工程、工业卫生、职业疾病、危险源辨识、事故调查以及相关行业安全（机械安全、电气安全、矿山安全、建筑安全、核安全、消防安全等）等，专业面广，涉及了本学科大量的专业词汇。每个单元都有课文生词和一些短语的注解。同时，本书还介绍了科技英语的特点，较为系统和详细地讲解了科技英语的翻译技巧，简单介绍了科技论文英文摘要和结论的写作要点，列举了大量的例句，有利于理解和掌握。本书无论是在内容选材上还是在内容编写上都具有专业特色和学术价值，实用性突出。

本书主要作为安全科学与工程类及其相关专业的本科教材，也可供安全科学与工程类专业的研究生以及从事安全技术与管理的专业人员学习参考。

图书在版编目（CIP）数据

安全工程专业英语/司鹄主编 . — 2 版 . — 北京：机械工业出版社，2018.7（2024.1重印）
高等教育安全科学与工程类系列教材
ISBN 978-7-111-59946-3

Ⅰ.①安…　Ⅱ.①司…　Ⅲ.①安全工程–英语–高等学校–教材　Ⅳ.①X93

中国版本图书馆 CIP 数据核字（2018）第 099026 号

机械工业出版社（北京市百万庄大街22号　邮政编码100037）
策划编辑：冷　彬　责任编辑：冷　彬　杨　洋　李　帅
责任校对：李　伟　封面设计：张　静
责任印制：张　博
北京建宏印刷有限公司
2024 年 1 月第 2 版第 7 次印刷
184mm×260mm · 11.75 印张 · 278 千字
标准书号：ISBN 978-7-111-59946-3
定价：39.00元

电话服务　　　　　　　　网络服务
客服电话：010-88361066　机　工　官　网：www.cmpbook.com
　　　　　010-88379833　机　工　官　博：weibo.com/cmp1952
　　　　　010-68326294　金　书　网：www.golden-book.com
封底无防伪标均为盗版　机工教育服务网：www.cmpedu.com

安全科学与工程类专业教材编审委员会

主 任 委 员：冯长根
副主任委员：王新泉　吴　超　蒋军成
秘 书 长：冷　彬
委　　　员：（排名不分先后）

冯长根　王新泉　吴　超　蒋军成　沈斐敏
钮英建　霍　然　孙　熙　王保国　王述洋
刘英学　金龙哲　张俭让　司　鹄　王凯全
董文庚　景国勋　柴建设　周长春　冷　彬

序[一]

"安全工程"本科专业是在1958年建立的"工业安全技术""工业卫生技术"和1983年建立的"矿山通风与安全"本科专业基础上发展起来的。1984年,国家教委将"安全工程"专业作为试办专业列入普通高等学校本科专业目录之中。1998年7月6日,教育部发文颁布《普通高等学校本科专业目录》,"安全工程"本科专业(代号:081002)属于工学门类的"环境与安全类"(代号:0810)学科下的两个专业之一[二]。据"高等学校安全工程学科教学指导委员会"1997年的调查结果显示,1958—1996年年底,全国各高校累计培养安全工程专业本科生8130人。到2005年年底,在教育部备案的设有安全工程本科专业的高校已达75所,2005年全国安全工程专业本科招生人数近3900名[三]。

按照《普通高等学校本科专业目录》的要求,以及院校招生和专业发展的需要,原来已设有与"安全工程"专业相近但专业名称有所差异的高校,现也大都更名为"安全工程"专业。专业名称统一后的"安全工程"专业,专业覆盖面大大拓宽[三]。同时,随着经济社会发展对安全工程专业人才要求的更新,安全工程专业的内涵也发生了很大变化,相应的专业培养目标、培养要求、主干学科、主要课程、主要实践性教学环节等都有了不同程度的变化,学生毕业后的执业身份是注册安全工程师。但是,安全工程专业的教材建设与专业的发展出现了不适应的新情况,无法满足和适应高等教育培养人才的需要。为此,组织编写、出版一套新的安全工程专业系列教材已成为众多院校的翘首之盼。

机械工业出版社是有着悠久历史的国家级优秀出版社,在高等学校安全工程学科教学指导委员会的指导和支持下,根据当前安全工程专业教育的发展现状,本着"大安全"的教育思想,进行了大量的调查研究工作,聘请了安全科学与工程领域一批学术造诣深、实践经验丰富的教授和专家,组织成立了教材编审委员会(以下简称"编审委"),决定组织编写"高等教育安全工程系列'十一五'教材"[四]。并先后于2004年8月(衡阳)、2005年8月(葫芦岛)、2005年12月(北京)、2006年4月(福州)组织召开了一系列安全工程专业本科教材建设研讨会,就安全工程专业本科教育的课程体系、课程教学内

[一] 此序作于2006年5月,为便于读者了解本套系列教材的产生与延续,该序将一直被保留和使用,并对其中某些的数据变化加以备注,以反映本套系列教材的可持续性,做到传承有序。

[二] 按《普通高等学校本科专业目录》(2012版),"安全工程"本科专业(专业代码:082901)属于工学学科的"安全科学与工程类"(专业代码:0829)下的专业。

[三] 这是安全工程本科专业发展过程中的一个历史数据,没有变更为当前数据是考虑到该专业每年的全国招生数量是变数,读者欲加了解,可在具有权威性的相关官方网站查得。

[四] 自2012年更名为"高等教育安全科学与工程类系列教材"。

容、教材建设等问题反复进行了研讨,在总结以往教学改革、教材编写经验的基础上,以推动安全工程专业教学改革和教材建设为宗旨,进行顶层设计,制订总体规划、出版进度和编写原则,计划分期分批出版30余门课程的教材,以尽快满足全国众多院校的教学需要,以后再根据专业方向的需要逐步增补。

由安全学原理、安全系统工程、安全人机工程学、安全管理学等课程构成的学科基础平台课程,已被安全科学与工程领域的学者认可并达成共识。本套系列教材编写、出版的基本思路是,在学科基础平台上,构建支撑安全工程专业的工程学原理与由关键性的主体技术组成的专业技术平台课程体系,编写、出版系列教材来支撑这个体系。

本套系列教材体系设计的原则是,重基本理论,重学科发展,理论联系实际,结合学生现状,体现人才培养要求。为保证教材的编写质量,本着"主编负责,主审把关"的原则,编审委组织专家分别对各门课程教材的编写大纲进行认真仔细的评审。教材初稿完成后又组织同行专家对书稿进行研讨,编者数易其稿,经反复推敲定稿后才最终进入出版流程。

作为一套全新的安全工程专业系列教材,其"新"主要体现在以下几点:

体系新。本套系列教材从"大安全"的专业要求出发,从整体上考虑、构建支撑安全工程学科专业技术平台的课程体系和各门课程的内容安排,按照教学改革方向要求的学时,统一协调与整合,形成一个完整的、各门课程之间有机联系的系列教材体系。

内容新。本套系列教材的突出特点是内容体系上的创新。它既注重知识的系统性、完整性,又特别注意各门学科基础平台课之间的关联,更注意后续的各门专业技术课与先修的学科基础平台课的衔接,充分考虑了安全工程学科知识体系的连贯性和各门课程教材间知识点的衔接、交叉和融合问题,努力消除相互关联课程中内容重复的现象,突出安全工程学科的工程学原理与关键性的主体技术,有利于学生的知识和技能的发展,有利于教学改革。

知识新。本套系列教材的主编大多由长期从事安全工程专业本科教学的教授担任,他们一直处于教学和科研的第一线,学术造诣深厚,教学经验丰富。在编写教材时,他们十分重视理论联系实际,注重引入新理论、新知识、新技术、新方法、新材料、新装备、新法规等理论研究、工程技术实践成果和各校教学改革的阶段性成果,充实与更新了知识点,增加了部分学科前沿方面的内容,充分体现了教材的先进性和前瞻性,以适应时代对安全工程高级专业技术人才的培育要求。本套系列教材中凡涉及安全生产的法律法规、技术标准、行业规范,全部采用最新颁布的版本。

安全是人类最重要和最基本的需求,是人民生命与健康的基本保障。一切生活、生产活动都源于生命的存在。如果人们失去了生命,一切都无从谈起。全世界平均每天发生约68.5万起事故,造成约2200人死亡的事实,使我们确认,安全不是别的什么,安全就是生命。安全生产是社会文明和进步的重要标志,是经济社会发展的综合反映,是落实以人为本的科学发展观的重要实践,是构建和谐社会的有力保障,是全面建成小康社会、统筹经济社会全面发展的重要内容,是实施可持续发展战略的组成部分,是各级政府履行市场监管和社会管理职能的基本任务,是企业生存、发展的基本要求。国内外实践证明,安全生产具有全局性、社会性、长期性、复杂性、科学性和规律性的特点,随着社会的不断进步,工业化进程的加快,安全生产工作的内涵发生了重大变化,它突破了时间和空间的限制，存在于人们日常生活和生产活动的全过程中，成为一个复杂多变的社会问题在安全领域的集中反映。安全问题不仅对生命个体非常重要，而且对社会稳定和经济发展产生重要影响。党的十六届五中全会提出"安全发展"的重要战略理念。安全发展是科学发展观理论体系的重要组成部分，安全发展与构建和谐社会有着密切的内在联系，以人为本，首先就是要以人的生命为本。"安全·生命·稳定·发展"是一个良性循环。安全科技工作者在促进、保证这一良性循环中起着重要作用。安全科技人才匮乏是我国安全生产形势严峻的重要原因之一。加快培养安全科技人才也是解开安全难题的钥匙之一。

高等院校安全工程专业是培养现代安全科学技术人才的基地。我深信，本套系列教材的出版，将对我国安全工程本科教育的发展和高级安全工程专业人才的培养起到十分积极的推进作用，同时，也为安全生产领域众多实际工作者提高专业理论水平提供学习资料。当然，这是第一套基于专业技术平台课程体系的教材，尽管我们的编审者、出版者夙兴夜寐，尽心竭力，但由于安全工程学科具有在理论上的综合性与应用上的广泛性相交叉的特性，开办安全工程专业的高等院校所依托的行业类型又涉及军工、航空、化工、石油、矿业、土木、交通、能源、环境、经济等诸多领域，安全科学与工程的应用也涉及人类生产、生活和生存的各个方面，因此本套系列教材依然会存在这样和那样的缺点、不足，难免挂一漏万。诚恳地希望得到有关专家、学者的关心与支持，希望选用本套系列教材的广大师生在使用过程中给我们多提意见和建议。谨祝本系列教材在编者、出版者、授课教师和学生的共同努力下，通过教学实践，获得进一步的完善和提高。

"嘤其鸣矣，求其友声"，高等院校安全工程专业正面临着前所未有的发展机遇，在此我们祝愿各个高校的安全工程专业越办越好，办出特色，为我国安全生产战线输送更多的优秀人才。让我们共同努力，为我国安全工程教育事业的发展做出贡献。

中国科学技术协会书记处书记[一]
中国职业安全健康协会副理事长
中国灾害防御协会副会长
亚洲安全工程学会主席
高等学校安全工程学科教学指导委员会副主任
安全科学与工程类专业教材编审委员会主任
北京理工大学教授、博士生导师

冯长根

[一] 曾任中国科协副主席。

前　言

随着科学技术的迅猛发展和经济的快速增长，安全问题越来越受到社会的关注。提高安全质量，是保持社会安定、促进经济可持续发展、改善人民生活水平的基础。

当今国际交流日益增多，要求安全科学与工程专业的人才具备扎实的专业知识，还要有良好的英语能力。本书正是为了满足高校安全科学与工程及相关专业学生的培养以及企业安全技术与管理人才的培训需要而编写的。

本书在借鉴国内外同类专著、教材的基础上编写而成。全书有较强的整体性和系统性，可以让读者对安全学科的认识和理解更为明确。同时，本书对科技英语的特点、翻译技巧以及英文摘要与结论的书写等内容的系统介绍，可以使读者对科技英语的认识更为清晰，有利于提高读者的英语阅读和写作能力。在此次第2版修订的过程中，编者根据教学反馈和安全科学与工程专业的课程设置情况，调整了部分单元的相关内容，突出体现了行业特色，满足当前教学对学生知识结构和能力的时代需要。

本书由重庆大学资源及环境科学学院安全科学与工程系司鹄教授主编；重庆大学资源及环境科学学院安全科学与工程系张东明教授、谢波副教授，重庆大学航空航天学院严波教授参加编写。其中，第1、2、3、4、6、7、8、9、10、16单元的英文课文以及科技英语摘要与结论的写作要点由司鹄编写，第11、12、13单元的英文课文由张东明编写，第5、14、15单元的英文课文由谢波编写，科技英语文章的特点和科技英语翻译技巧由司鹄和严波合作编写。

全书由司鹄负责统稿。中南大学资源与安全工程学院的吴超教授担任主审，对本书进行了全面、认真、严格、细致的审查，提出了许多宝贵的修改意见和建议。中原工学院的王新泉教授也对本书的编写提出了宝贵的建议。本书的编审工作是在安全科学与工程类专业教材编审委员会的指导下进行的，其大纲的编制、审定以及相关内容的取舍，均经过编审委员会的反复讨论定夺，而且编审委员会也曾多次组织专家对书稿进行审稿工作。在此，对上述专家、同仁的辛勤工作表示衷心的感谢。

本书的编写参考了国内外安全管理学、安全系统工程、安全人机工程、职业安全与健康管理等相关文章和有关书籍，在此，谨对原作者和研究者表示最诚挚的谢意。

编写本书的过程也是一个不断学习、不断提高的过程。由于编者水平有限，书中难免有不妥与错误之处，敬请广大读者及相关专家批评指正。

编　者

目 录

序

前言

Unit One　Safety Management Systems ……………………………………………………… 1

　　Translation Skill　科技文章的特点 ………………………………………………………… 5

　　Reading Material　Integrated, Incident-Wide Safety Management ……………………… 7

Unit Two　System Safety Engineering …………………………………………………… 11

　　Translation Skill　科技英语翻译技巧（一）——词义引申 …………………………… 15

　　Reading Material　Basics of Safety Engineering ………………………………………… 16

Unit Three　The Ergonomics Process …………………………………………………… 21

　　Translation Skill　科技英语翻译技巧（二）——词量增减 …………………………… 25

　　Reading Material　Implementation of Human Error Diagnosis System ……………… 26

Unit Four　Hazard Identification ………………………………………………………… 31

　　Translation Skill　科技英语翻译技巧（三）——词性转换 …………………………… 35

　　Reading Material　Analyzing Hazards …………………………………………………… 36

Unit Five　What Is an OHSMS? ………………………………………………………… 41

　　Translation Skill　科技英语翻译技巧（四）——句子成分转换 ……………………… 45

　　Reading Material　The Standard for Occupational Health and Safety ………………… 46

Unit Six　Industrial Hygiene ……………………………………………………………… 50

　　Translation Skill　科技英语翻译技巧（五）——常见多功能词的译法（Ⅰ） ………… 55

　　Reading Material　Occupational Illness …………………………………………………… 57

Unit Seven　Safety Culture ………………………………………………………………… 63

　　Translation Skill　科技英语翻译技巧（六）——常见多功能词的译法（Ⅱ） ………… 67

　　Reading Material　Perspectives on Safety Culture ……………………………………… 69

目 录

Unit Eight　Motivating Safety and Health ··· 75
　Translation Skill　科技英语翻译技巧（七）——数词的译法 ·············· 79
　Reading Material　The Motivational Environment ·································· 81

Unit Nine　Accident Investigations ·· 88
　Translation Skill　科技英语翻译技巧（八）——被动语态的译法 ········ 92
　Reading Material　Cooperation between Insurance and Prevention ············ 94

Unit Ten　Safety Electricity ··· 98
　Translation Skill　科技英语翻译技巧（九）——定语从句及同位语从句的
　　　　　　　　　译法（I） ·· 103
　Reading Material　Physiological Effects of Electricity ···························· 104

Unit Eleven　Machinery Equipment Safety ····································· 110
　Translation Skill　科技英语翻译技巧（十）——定语从句及同位语从句的
　　　　　　　　　译法（II） ··· 114
　Reading Material　Machine Guarding ·· 116

Unit Twelve　Accident Analysis in Construction ······························ 121
　Translation Skill　科技英语翻译技巧（十一）——状语从句的译法 ····· 126
　Reading Material　Fall Prevention ·· 128

Unit Thirteen　Accident Analysis in Mine Industry ··························· 133
　Translation Skill　科技英语翻译技巧（十二）——长句的译法 ·········· 137
　Reading Material　Explosions in Gobs in Coal Mines ··························· 139

Unit Fourteen　Hazardous Chemical and Its Identication ···················· 142
　Translation Skill　科技应用文的译法 ··· 145
　Reading Material　Basic Principles for Controlling Chemical Hazards ········ 148

Unit Fifteen　Combustion and Explosion Accidents ··························· 152
　Writing Skill　科技英语摘要的写作要点 ··· 155
　Reading Material　Prevention and Protection for Dust Explosion ·············· 157

Unit Sixteen　The History of Nuclear Power Plant Safety ···················· 163
　Writing Skill　科技英语结论的写作要点 ··· 167
　Reading Material　Railway Safety Management ·································· 169

参考文献 ·· 174

Unit One

Safety Management Systems

1. Accident Causation Models

The most important aim of safety management is to maintain and promote workers' health and safety at work. Understanding why and how accidents and other unwanted events develop is important when preventive activities are planned. Accident theories aim to clarify the accident phenomena, and to explain the mechanisms that lead to accidents. All modern theories are based on accident causation models which try to explain the sequence of events that finally produce the loss. In ancient times, accidents were seen as an act of God and very little could be done to prevent them. In the beginning of the 20th century, it was believed that the poor physical conditions are the root causes of accidents. Safety practitioners concentrated on improving machine guarding, housekeeping, and inspections. In most cases an accident is the result of two things: the human act, and the condition of the physical or social environment.

Petersen extended the causation theory from the individual acts and local conditions to the management system. He concluded that unsafe acts, unsafe conditions, and accidents are all symptoms of something wrong in the organizational management system. Furthermore, he stated that it is the top management who is responsible for building up such a system that can effectively control the hazards associated to the organization's operation. The errors done by a single person can be intentional or unintentional. Rasmussen and Jensen have presented a three-level skill-rule-knowledge model for describing the origins of the different types of human errors. Nowadays, this model is one of the standard methods in the examination of human errors at work.

Accident-proneness models suggest that some people are more likely to suffer an accident than others. The first model was created in 1919, based on statistical examinations in a munitions factory. This model dominated the safety thinking and research for almost 50 years, and it is still used in some organizations. As a result of this thinking, accident was blamed solely on employees rather than the work process or poor management practices. Since investigations to discover the underlying causal factors were felt unnecessary and/or too costly, a little attention was paid to how accidents actually happened. Employees' attitudes towards risks and risk taking have been studied, e. g. by Sulzer-Azaroff. According to her, employees often behave unsafely, even when they are fully aware of the risks involved. Many research results also show that the traditional promotion methods like campaigns, posters and safety slogans have seldom increased the use of safe work practices. When backed up by other activities such as training, these measures have been somewhat more effective. Experiences on some successful methods to change employee behavior and attitudes have been reported. One well-known method is a small-group process used for improving housekeeping in industrial workplaces. A comprehensive model of accident causation has been presented by Reason who introduced the concept of organizational error. He stated that corporate culture is the starting-point of the accident sequence. Local conditions and human behavior are only contributing factors in the build-up of the undesired event. The latent organizational failures lead to accidents and incidents when penetrating system's defenses and barriers. Groeneweg has developed Reason's model by classifying the typical latent error types. His TRIPOD model calls the different errors as General Failure Types (GFTs). The concept of organizational error is in conjunction with the fact that some organizations behave more safely than others. It is often said that these organizations have good safety culture. After the Chernobyl accident, this term became well-known also to the public.

Loss prevention is a concept that is often used in the context of hazard control in process industry. Lees has pointed out that loss prevention differs from traditional safety approach in several ways. For example, there is more emphasis on foreseeing hazards and taking actions before accidents occur. Also, there is more emphasis on a systematic rather than a trial and error approach. This is also natural, since accidents in process industry can have catastrophic consequences. Besides the injuries to people, the damage to plant and loss of profit are major concerns in loss prevention. The future research on the ultimate causes of accidents seems to focus on the functioning and management of the organization. The strategic management, leadership, motivation, and the personnel's visible and hidden values are some issues that are now under intensive study.

2. Safety Management as an Organizational Activity

Safety management is one of the management activities of a company. Different companies have different management practices, and also different ways to control health and safety hazards. Organizational culture is a major component affecting organizational performance and behavior. One comprehensive definition for an organizational culture has been presented by Schein who has said that organizational culture is "a pattern of basic assumptions—invented, discovered, or developed by a given group as it learns to cope with its problems of external adaptation and internal integration—that has worked well enough to be considered valid and, therefore, to be taught to new members as the correct way to perceive, think, and feel in relation to those problems". The concept of safety culture today is under intensive study in industrialized countries. Booth & Lee have stated that an organization's safety culture is a subset of the overall organizational culture. This argument, in fact, suggests that a company's organizational culture also determines the maximum level of safety the company can reach. The safety culture of an organization is the product of individual and group values, attitudes, perceptions, competencies, and patterns of behavior that determine the commitment to, and the style and proficiency of, an organization's health and safety management. Furthermore, organizations with a positive safety culture are characterized by communications founded on mutual trust, by shared perceptions of the importance of safety, and by confidence in the efficacy of preventive measures. There have been many attempts to develop methods for measuring safety culture. Williamson et al. have summarized some of the factors that the various studies have shown to influence organization's safety culture. These include: organizational responsibility for safety, management attitudes towards safety, management activity in responding to health and safety problems, safety training and promotion, level of risk at the workplace, workers' involvement in safety, and status of the safety officer and the safety committee.

Organizations behave differently in the different parts of the world. This causes visible differences also in safety activities, both in employee level and in the management level. Reasons for these differences are discussed in the following. The studies of Wobbe reveal that shop-floor workers in the USA are, in general, less trained and less adaptable than those in Germany or Japan. Wobbe claims that one reason for this is that, in the USA, companies providing further training for their staff can expect to lose these people to the competitors. This is not so common in Europe or in Japan. Furthermore, for unionized companies in the USA, seniority is valued very highly, while training or individual's skills and qualifications do not effect job security, employment, and wage levels very much. Oxenburgh has studied the total costs of absence from work, and found that local culture and legislation has a strong effect on absenteeism rates. For example, the national systems for paying and receiving compensation explain the differences to some extent. Oxenburgh mentions Sweden as a high absenteeism country, and Australia as a low absenteeism country. In Sweden injuries and illnesses are paid by the state social security system, while in Australia, the employer pays all these costs, including illnesses not related to work. Comparison of accident statistics reveals that there are great national differences in accident frequencies and in the accident related absenteeism from work. Some of the differences can be explained by the different accident reporting systems. For example, in some countries only absenteeism lasting more than three working days is included in the statistics. The frequency of minor accidents varies a lot according to the possibility to arrange substitutive work to the injured worker. Placing the injured worker to another job or to training is a common practice for example in the USA and in the UK, while in the Scandinavian countries this is a rarely used procedure.

Some organizations are more aware of the importance of health and safety at work than others. Clear development stages can be found in the process of improving the management of safety. Waring has divided organizations to three classes according to their maturity and ability to create an effective safety management system. Waring calls the three organizational models as the mechanical model, the socio-technical model, and the human activity system approach. In the mechanical model, the structures and processes of an organization are well-defined and logical, but people as individuals, groups, and the whole organizations are not considered. The socio-technical model is an approach to work design which recognizes the interaction of technology and people, and which produces work systems that are technically effective and have characters that lead to high job satisfaction. A positive dimension in this model is that human factors are seen important, for example, in communication, training and emergency responses. The last model, the human activity system approach focuses on people, and points out the complexity of organizations. The strength of this approach is that both formal (or technical) paradigms and human aspects like motivation, learning, culture, and power relations are considered. Waring points out that although the human activity approach does not automatically guarantee success, it has proven to be beneficial to organizations in the long run.

3. Safety Policy and Planning

A status review is the basis for a safety policy and the planning of safety activities. According to BS 8800 a status review should compare the company's existing arrangements with the applicable legal requirements, organization's current safety guidelines, best practices in the industry's branch, and the existing resources directed to safety activities. A thorough review ensures that the safety policy and the activities are developed specifically according to the needs of the company.

A safety policy is the management's expression of the direction to be followed in the organization. According to Petersen, a safety policy should commit the management at all levels and it should indicate which tasks, responsibilities and decisions are left to lower-level management. Booth and Lee have stated that a safety policy should also include safety goals as well as quantified objectives and priorities. The standard BS 8800 suggests that in the safety policy, management should show commitment to the following subjects:

- Health and safety are recognized as an integral part of business performance.
- A high level of health and safety performance is a goal which is achieved by using the legal requirements as the minimum, and where the continual cost-effective improvement of performance is the way to do things.
- Adequate and appropriate resources are provided to implement the safety policy.
- The health and safety objectives are set and published at least by internal notification.
- The management of health and safety is a prime responsibility of the management, from the most senior executive to the supervisory level.
- The policy is understood, implemented, and maintained at all levels in the organization.
- Employees are involved and consulted in order to gain commitment to the policy and its implementation.
- The policy and the management system are reviewed periodically, and the compliance of the policy is audited on a regular basis.
- It is ensured that employees receive appropriate training, and are competent to carry out their duties and responsibilities.

Some companies have developed so-called "safety principles" which cover the key areas of the company's safety policy. These principles are utilized as safety guidelines that are easy to remember, and which are often placed on wall-boards and other public areas in the company. As an example, the DuPont company's safety principles are the following:

- All injuries and occupational illnesses can be prevented.
- Management is responsible for safety.
- Safety is an individual's responsibility and a condition of employment.
- Training is an essential element for safe workplaces.
- Audits must be conducted.
- All deficiencies must be corrected promptly.
- It is essential to investigate all injuries and incidents with injury potential.
- Off-the-job safety is an important part of the safety effort.
- It is good business to prevent injuries and illnesses.

- People are the most important element of the safety and occupational health program.

The safety policy should be put into practice through careful planning of the safety activities. Planning means determination of the safety objectives and priorities, and preparation of the working program to achieve the goals. A company can have different objectives and priorities according to the nature of the typical hazards, and the current status of hazard control. However, some common elements to a safety activity planning can be found. According to BS 8800, the plan should include:
- appropriate and adequately resourced arrangements, competent personnel who have defined responsibilities, and effective channels of communication;
- procedures to set objectives, device and implement plans to meet the objectives, and to monitor both the implementation and effectiveness of the plans;
- description of the hazard identification and assessment activities;
- methods and techniques for measuring safety performance, and in such way that absence of hazardous events is not seen as evidence that all is well.

In the Member States of the European Union, the "framework" Directive 89/391/EEC (European Economic Community) obligates the employer to prepare a safety program that defines how the effects of technology, work methods, working conditions, social relationships and work environment are controlled. According to Walters, this directive was originally passed to harmonize the overall safety strategies within the Member States, and to establish a common approach to the management and organization of safety at work. Planning of the safety activities is often done within the framework of quality and environmental management systems.

New Words and Expressions

preventive [prɪˈventɪv]	adj. 预防性的
proneness [ˈprəʊnɪs]	n. 俯伏；倾向
munition [mjuː(ː)ˈnɪʃən]	n. 军需品 v. 供给军需品
dominate [ˈdɒmɪneɪt]	v. 支配，占优势
blame sth. on sb.	把某事的责任归咎于某人 [事]
integration [ˌɪntɪˈgreɪʃən]	n. 综合
dimension [dɪˈmenʃən]	n. 尺寸，尺度；维（数）
paradigm [ˈpærədaɪm, -dɪm]	n. 范例
promptly [ˈprɒmptli]	adv. 敏捷地，迅速地
EEC (European Economic Community)	欧洲经济共同体

Translation Skill

科技文章的特点

科技文章的特点是：清晰、准确、精炼、严密。现将科技文章的语言结构特色陈述如下。

一、大量使用名词化结构

科技英语要求行文简洁、表达客观、内容准确、信息量大，常强调存在的事实，而非某一行为，所以大量使用名词化结构。

The earth rotates on its own axis, which causes the change from day to night.

The rotation of the earth on its own axis causes the change from day to night.

名词化结构 The rotation of the earth on its own axis 使复合句简化成简单句，并且使表达的概念更加确切、严密。

科技英语所表达的是客观规律，应尽量避免使用第一、第二人称；此外，在科技英语的表达中，常常将主要信息置于句首。

The vision of health, safety, environment responsibility and company values demonstrate health and safety in the workplace is fundamental.

二、广泛使用被动句

科技文章侧重叙事、推理，强调客观、准确，因而大量采用第三人称叙述，使用被动语态。过多使用第一、第二人称，会造成主观臆断的现象。科技英语中的谓语至少三分之一是被动语态。

The safety policy should be put into practice through careful planning of the safety activities.

Occupational safety and health has received increasing attention due to its undeniable influence on economic development and social stability.

三、非限定动词

由于科技文章要求书写简练、结构紧凑，因而常常使用分词短语代替定语从句；使用分词独立结构代替状语从句或并列分句；使用不定式代替各种从句，"介词 + 动名词短语"代替定语从句或状语从句。这样，既可缩短句子，又比较醒目。

Organizational culture is a major component *affecting* organizational performance and behavior.

A safety policy is the management's expression of the direction *to be followed* in the organization.

The most important aim of safety management is *to maintain and promote* workers' health and safety at work.

四、后置定语

大量使用后置定语是科技文章的特点之一。常用的句子结构形式有：介词短语后置、形容词及形容词短语后置、副词后置、定语从句后置。

In small and medium-sized companies, the safety manager and the safety representative often have other duties *besides their health and safety tasks*. (介词短语后置)

The safety manager's role is to act as an expert *who is aware of the health and safety legislation and other obligations concerning the company*. (定语从句后置)

The efforts *necessary to assure that sufficient emphasis is placed on system safety* are often organized into formal programs. ("形容词短语 + 从句"后置)

五、长句

为了表达一个复杂概念，使之逻辑严密，科技文章中常常出现许多长句。

One comprehensive definition for an organizational culture has been presented by Schein who has said that organizational culture is "a pattern of basic assumptions—invented, discovered, or developed by a given group as it learns to cope with its problems of external adaptation and internal integration—that has worked well enough to be considered valid and, therefore, to be taught to new members as the correct way to perceive, think, and feel in relation to those problems".

Reading Material

Integrated, Incident-Wide Safety Management

In developing recommendations to improve safety management during the response to a major disaster, providing better ways for individual response organizations to gather information, to analyze risk and make decisions, and to take action would not be enough to fully address the safety management needs during large-scale operations. Rather, the complexity and demands of post-disaster environments call for solutions based on improved coordination among the multiple organizations that become involved in major disaster response operations.

Nothing demonstrated this better than the response operations at the Pentagon and World Trade Center on September 11, 2001. What we learned from those examples led us to the central organizational finding of this study: The emergency response community should put in place structures and preparedness efforts that will formalize an integrated, incident-wide approach to safety management at major disaster response operations. Indeed, the solutions to key problems in each functional phase of the safety management cycle are inherently inter-organizational, relying on multi-agency safety efforts:

(1) **Gathering Information**

Required hazard monitoring capabilities may reside in different response organizations.

Information on responder accountability, training, equipment, and health status information must come from many separate organizations.

(2) **Analyzing Options and Making Decisions**

Technical expertise to assess hazards must frequently be drawn from multiple responding organizations.

Effective decision-making requires coordination of equipment and hazard mitigation options brought to the incident by all responding organizations.

(3) **Taking Action**

Difficulties in uniform safety enforcement can be addressed only via interagency coordination and agreement.

Sustainability measures to protect responder health must be applied across organizational boundaries.

Management of human and material safety resources must be coordinated among multiple responding organizations.

Only by building the capability of response units and agencies to coordinate at the organizational level can they be most prepared to successfully manage the functional challenges they face.

Developing such an integrated approach requires a transition from viewing safety management as an activity primarily carried out by individual organizations alone to understanding it as a multi-agency function within the ICS (Incident Command System) that can scale up to meet the needs of complex disaster response operations. This transition must encompass organizations across the full range of the disaster response community—all levels of government, nongovernmental groups, and the private sector. In addition, recognizing the high-pressure and severely time-constrained post-disaster environment, this functional approach to safety must facilitate rapid initiation of multi-agency coordination and safety management activities.

1. Benefits of an Integrated, Incident-Wide Safety Management Approach

The capability to draw on the safety resources of many organizations and effectively apply them to safety management for the overall incident would provide several important opportunities to better meet the safety needs of all involved responders:
- access to the specialized safety capabilities of multiple organizations;
- a strategic approach to safety management;
- a mechanism to address inherently multi-agency safety issues;
- a route to take advantage of diverse response capabilities.

2. Access to the Specialized Safety Capabilities of Multiple Organizations

When organizations from different response disciplines come together at major disaster operations, they bring significantly different levels of safety management capability. Such differences in expertise and equipment can result in safety shortfalls when organizations without necessary expertise or equipment are "on their own" to manage responder safety. However, when safety management efforts are coordinated among multiple agencies, such differences represent an opportunity to draw on organizations' relative strengths to bolster protection for responders overall.

Many of the different organizations involved in carrying out response tasks at an incident scene bring not only operational capabilities, but safety expertise and resources to the operation. Government agencies at all levels, nongovernmental organizations, and private-sector entities with safety-related responsibilities at the scene may bring additional safety resources and knowledge. Examples include:
- law enforcement and intelligence expertise on potential threats and security hazards after terrorist events;
- fire department expertise with thermal hazards and hazardous materials operations;
- public health organizations' capabilities in disease surveillance and health monitoring;
- departments of Defense and Energy expertise on nuclear, radiological, and other weapons of mass destruction;
- utility, transportation, or construction capabilities in their areas of specialization and responsibility;
- federal, state, local or other organizations' expertise to assess hazards and measure environmental and occupational exposures.

It would be impractical for individual organizations to maintain the equipment and expertise needed to cope with all the hazards that could arise during a response to a major disaster. An integrated, incident-wide approach to safety makes better safety management resources accessible than would be possible for organizations operating alone.

3. A Strategic Approach to Safety Management

Just as the Incident Commander needs to take a strategic viewpoint of a disaster operation, a safety manager must be able to consider safety needs from an overall, strategic perspective. If the individuals responsible for managing responder safety are too close to or absorbed in the details of an operation, it is much less likely that they will be able to fully understand and address the risks at a complex disaster scene. This can make it difficult or impossible to make good safety decisions and meet worker safety needs. Similarly, if safety managers cannot take a long-term view of safety concerns—for example, anticipating response safety concerns and projecting safety requirements—safety management will also suffer.

For the safety manager of an individual organization, the complexity and operational demands of a major disaster make it exceedingly difficult to get this overall perspective or to project future safety needs. But in the context of an integrated approach, the additional expertise, capabilities, and resources that can be brought to bear on safety issues can help build and maintain this more strategic approach to the incident. By delegating specific tasks—such as technical monitoring of hazards, equipment logistics, or accounting for personnel—to the right experts or organizations, safety managers can focus their attention on building an overall understanding of the incident safety needs, providing better support to the incident commander on the safety components of operational decisions, and anticipating safety and health concerns that may arise as the incident evolves.

4. A Mechanism to Address Inherently Multi-agency Safety Issues

A coordinated safety management effort provides a mechanism for sharing necessary safety information among response organizations. This coordination is particularly important to address the possibility that response activities can produce new and unfamiliar safety hazards for other responders. Similarly, integrating multi-agency activities can improve the effectiveness of safety measures by allowing better coordination of safety logistics efforts. Such integration would reduce the chance of duplicative resource requests from separate organizations, a situation often observed in major disaster responses, and potentially make it possible to better allocate safety resources across the response overall.

An integrated approach to safety management can also make it possible to begin addressing a potentially more serious problem—the difficulty in uniformly implementing and, if necessary, enforcing safety policies across the disaster response operation. By bringing together representatives from relevant organizations, integrating different organizations' safety management efforts provides a route to build consensus on safety policies and procedures among all response organizations. Such an incident-level consensus would enable more uniform implementation of safety measures across an incident, even in the absence of centralized safety enforcement authority. If incident-wide enforcement measures become necessary to ensure use of critical safety measures, an integrated approach provides a way to develop the necessary multi-agency commitment to put them in place.

5. A Route to Take Advantage of Diverse Response Capabilities

Responders from different disciplines come to an event with unique types of expertise. In addition, organizations that more frequently face particular types of disasters—for example, responders from areas that experience specific natural disasters—develop expertise in responding to those sorts of incidents. Specialized expertise may also reside in response organizations from areas with elevated risk of particular events—such as high-profile cities at higher risk of terrorist attack—because of increased preparedness or participation in exercises aimed at those events. Accordingly, particular response units may be significantly more qualified to operate safely in particular risk environments. An integrated approach to safety management permits decision-makers to draw upon this diversity to ensure that responders are assigned those tasks they are especially qualified and equipped to perform safely, lowering the safety risks for other responders.

6. Implementing an Integrated, Incident-Wide Safety Management

Responders to recent large-scale disasters have recognized the need to integrate their efforts in order to address the complex safety concerns of emergency workers. At both the Pentagon and World Trade Center, the practical difficulties associated with managing responder safety led response organizations to implement ad hoc arrangements to coordinate their safety efforts. Responders at the World Trade Center, for example, formed a large safety team, held daily safety-focused meetings, and brought safety experts into incident command meetings. This safety team initially instituted an accident prevention plan for the site and eventually developed a comprehensive safety and health management plan with input from the four primary contractors and 26 federal, state, and local agencies operating at the Trade Center site.

Although these ad hoc efforts broke important ground by recognizing the need to implement an integrated, incident-wide approach to safety management, they also had significant shortcomings. First, because these expedient arrangements were developed during the course of the response, they took time to put in place. During the days before the structures were set up, the safety efforts of responding organizations had no effective mechanism for integration. In general, depending on the specific hazards involved in an incident, such delays could have significant consequences for the safety of responders. Second, improvised groups also may overlook the involvement of important, but less obvious, sources of expertise needed for managing responder safety and health. For example, it was not always fully clear to responders at the World Trade Center disaster how the participants in the safety meetings were determined. It sometimes seemed to require significant "negotiation" to gain access to the meetings.

In addition, some responders perceived it as a weakness that these safety management structures existed outside the formal ICS. Interviewees indicated that it was not always clear how effectively the deliberations of the safety committee were connected with the ICS. When safety is managed by an ad hoc group, one interviewee commented, it is less clear "how decisions are actually being made", and both the perceived validity of the decisions and accountability of the decision-makers can be weakened.

New Words and Expressions

preparedness [prɪˈpeərɪdnɪs]　　　　　n. 有准备, 已准备
incident [ˈɪnsɪdənt]　　　　　　　　　n. 事件, 事变
multiple [ˈmʌltɪpl]　　　　　　　　　adj. 多样的, 多重的　n. 倍数, 若干
logistics [ləʊˈdʒɪstɪks]　　　　　　　n. 物流, 后勤学; 后勤
ad hoc　　　　　　　　　　　　　　adj., adv. [拉] 特别的 [地]; 尤其, 关于这; 非正式的; 特定的
expedient [ɪksˈpiːdɪənt]　　　　　　　adj. 有利的　n. 权宜之计

Unit Two

System Safety Engineering

1. System Safety

Safety has been intentionally integrated into the design and development of many systems. In some instances, however, the safety realized in a system was only a byproduct of good engineering practice. As systems have become more complex and expensive, safety problems have become more acute. Because of accelerating technology and demands for a "first time safe" operation, a need has arisen to formally organize safety efforts throughout a system's life cycle. To meet this need, the concepts of system safety and the field of system safety engineering have evolved.

System safety concepts are based on the idea that an optimum degree of safety can be achieved within the constraints of system effectiveness. This optimum is attained through a logical reasoning process. Accidents, or potential accidents, are first considered to be the result of a number of interacting causes within the system. From a deliberate analysis of the system, each accident cause and interaction is logically identified and evaluated. Work may then be performed to eliminate or otherwise control these accident causes.

System safety engineering draws upon system safety concepts and the approaches used in system engineering. Its objective is to safely integrate all system components in a manner consistent with other system criteria. It involves the application of scientific and management principles for the timely recognition, evaluation, and control of hazards. Through the logical programming of these efforts over a system's life cycle, the desired level of safety can be realized.

2. System Safety Program

The degree of safety achieved in a system depends on the importance it is given by system designers, managers, and operators. The efforts necessary to assure that sufficient emphasis is placed on system safety are often organized into formal programs. The objectives of such programs are to recognize, evaluate, and control system hazards as early in the life cycle as possible. Adequate safety input during the initial phases is the key to produce an inherently safe system. Effective system safety programs also eliminate the schedule delays and costly changes that often result in systems that do not have adequate safety planning.

(1) **Concept Phase**

Program activities in the concept phase include establishing safety criteria within the definition of the system task. These criteria normally provide for compliance with accepted safety standards such as those contained in government regulations. The maximum risks that are acceptable in system operation may also be specified in the task definition. A gross hazard analysis is often performed for each alternative approach to system design. The results of these analyses highlight special areas for safety consideration and are useful in evaluating each alternative. Knowledge of the safety experience of related systems will also aid in selecting the best alternative. In addition, program requirements and schedules for the remainder of the life cycle are established in the concept phase.

(2) **Design Phase**

Equipment specifications, maintenance plans, training plans, proposed system procedures, and other design materials are reviewed and evaluated during the design phase. Extensive hazard analyses are also conducted to identify and resolve potential safety problems. All system components and interactions are studied to predict the risks that will be involved in system operation. Design changes are then made to minimize these risks. Knowledge of similar systems is also helpful in this phase. Of particular interest are measures taken to correct design features that resulted in accidents. Similar accidents may be avoided by including the same corrective measures in the specifications that emerge from the design phase.

(3) **Development Phase**

During the development phase, system safety programs assure that the design requirements are incorporated as the system is generated. Hazard analyses are performed or updated. Additional studies or tests are conducted to verify the adequacy of safety design features. Training courses for operating personnel are reviewed to confirm that there is satisfactory coverage of safety aspects. Refinements of system design often result from these evaluations to improve system safety prior to the operation phase.

(4) **Operation Phase**

Periodic safety inspections, maintenance, training, and performance reviews are conducted in the operation phase to maintain or improve the safety achieved previously. Hazard analyses are updated to determine the impact of any system changes, and action is taken to control any hazards that result from such changes. Similar action is taken following any accident or malfunction to correct deficiencies that have been overlooked.

(5) **Disposal Phase**

Program efforts are continued through the final phase of the life cycle. System disposal procedures are reviewed and monitored. Special precautions are taken in disposing of hazardous materials, equipment, etc.

3. Hazard Analysis

The heart of a system safety program is in the performance of hazard analyses. These efforts also involve most of the detail work of system safety engineering. Analyses are conducted to identify and evaluate hazards within a system. With this information, responsible officials may determine the safest, most efficient means of controlling the hazards identified. With its accident sources eliminated, the entire system also becomes

more effective in performing its task. In general, hazard analyses are conducted as follows:
- gain an understanding of the system;
- define the scope and purpose of the analysis;
- select and apply an analysis technique;
- evaluate the results.

Hazard analysis is not an intuitive process. For an analysis to be meaningful, it must be logical, accurate, descriptive of the system, and based on valid assumptions. Its success largely depends on the skill and knowledge of those conducting the analysis. Anyone who has a thorough, working knowledge of both the system under consideration and the analysis technique to be used may perform a hazard analysis. In practice, the efforts of several persons with varying backgrounds are usually required to assure that meaningful and comprehensive hazard information is obtained.

A system safety program usually requires the performance of several hazard analyses throughout the life cycle. The scope and purpose of each required analysis may be different. For example, a preliminary hazard analysis may be required in the concept phase to provide an initial assessment of hazards that may be encountered in subsequent life cycle phases. Special analyses may also be specified to determine the hazards involved in system development, assembly, operation, maintenance, or disposal. In addition, the entire system or only part of the system may be analyzed at one time. Where subsystem hazard analyses are performed, a total system analysis should also be conducted to identify potential hazards at the interfaces of the subsystems.

Regardless of the scope, purpose, or technique used, a hazard analysis is only a tool. It does not make decisions. Analysis was never intended to provide the final answer to system safety problems, but rather to aid in making decisions for improving the system.

4. Hazard Control

To control the hazards identified by analysis, the system or its environment must be altered. Hazard control typically is accomplished through either engineering, educational, or administrative solutions. Where these types of hazard controls are insufficient, protective apparel is often used.

5. Engineering Solutions

Because engineering solutions are relatively permanent, they are the most desirable type of hazard control. They normally involve altering the system's machines, materials, or environment. The approach used in engineering solutions is to minimize hazards through design. Where possible, a system design that completely eliminates an identified hazard should be selected. A hazard that cannot be eliminated through design selection should be controlled by the use of safety design features such as fixed or automatic protective devices. When neither design selection nor design features can effectively control a known hazard, warning devices should be used to detect the hazardous condition and generate a warning signal. Such signals should be standardized and designed to minimize the probability of incorrect interpretation by operating personnel.

Of primary importance in engineering solutions is reducing the amount of energy or hazardous material in the system. Where the desired reduction is not possible, personnel should be protected from the source of energy or hazardous material. Equipment such as shields, barricades, cabs, and canopies are often used to protect people and property, should an unplanned release of energy or hazardous material occur. Changes in tools, lighting, atmospheric conditions, work area layout, and job location are other examples of engineering solutions.

6. Educational Solutions

Hazards involving the behavior of the people within a system are often controlled through educational solutions. Suitable education and training programs are instituted as a means of modifying behavior and thus improving system effectiveness. These programs include formal training sessions, periodic safety meetings, safety promotion programs, etc.

7. Administrative Solutions

Administrative solutions often involve changing the method or procedures followed in performing specific tasks within a system to reduce accident potential. In addition, changes in the procedures for personnel selection, assignment, and training, and those for system housekeeping, maintenance, inspection, etc., are included in the category of administrative solutions.

8. Protective Apparel

At times, a hazard cannot be controlled to an acceptable degree through engineering, educational, or administrative means. For these hazards, the best course of action is to use protective apparel to protect the worker in the event of a mishap. Included here are those safety devices the worker would wear on his or her person such as special clothing, safety shoes, goggles, safety belts, gloves, respirators, safety glasses, etc.

9. Summary

System safety engineering is a relatively new approach to accident prevention. Its concepts and techniques have evolved from efforts to improve the safety of the complex technical systems that are common in today's society. It is based on the ideas that accidents result from a number of interacting causes within a system, and that each cause and interaction can be logically identified, evaluated, and controlled. Through the logical application of scientific and management principles over the life cycle of a system, system safety engineering attempts to achieve an optimum degree of safety.

The efforts necessary to achieve the desired degree of safety are usually organized into formal programs. The objective of such programs is to assure that system hazards are eliminated or otherwise controlled as early in the life cycle as possible. Most of the detail work involved in a system safety program is in the performance of hazard analyses. With the information provided by analysis, responsible officials can determine the safest, most efficient means of controlling the hazards identified.

New Words and Expressions

malfunction [mælˈfʌŋkʃən]　　　　n. 故障
apparel [əˈpærəl]　　　　　　　　n. 衣服；装饰
mishap [ˈmɪshæp]　　　　　　　　n. 灾祸
goggle [ˈgɒgl]　　　　　　　　　n. 风镜，护目镜

Translation Skill

科技英语翻译技巧（一）——词义引申

所谓词义的引申，是指在一个词所具有的基本词义的基础上，进一步地加以引申，选择比较恰当的汉语来表达，使原文的思想表现得更加准确，译文更加通顺、流畅。词义的引申主要有词义转译、词义具体化、词义抽象化和词的搭配4种手段。

一、词义转译

遇到一些无法直译或不宜直译的词或词组时，应根据上下文和逻辑关系引申转译。

Like any precision device, the monitor of methane requires careful *treatment*.
像任何精密仪器一样，瓦斯监测仪也需要精心维护。（不译成"待遇"）

Solar energy seems to *offer more hope* than any other source of energy.
太阳能似乎比其他能源更具有前景。（不译成"提供更多希望"）

二、词义具体化

有时应根据汉语的表达习惯，把原文中某些词义较笼统的词引申为词义较具体的词。

The purpose of a driller is *to holes*.
钻机的功能是钻孔。

A single-point cutting tool is used *to cut* threads on engine lathes.
普通车床是用单刃刀具来车螺纹的。

三、词义抽象化

把原文中词义较具体的词引申为词义较抽象的词，或把词义较形象的词引申为词义较一般的词。

The major contributors in component technology have been the semi-conductor components.
元件技术中起主要作用的是半导体元件。（不译成"主要贡献者"）

There are three steps which must be taken before we *graduate from* the integrated circuit technology.
我们要完全掌握集成电路技术，还必须经过三个阶段。（不译成"毕业于"）

四、词的搭配

遇到动词、形容词与名词搭配时，应根据汉语的搭配习惯，而不应受原文字面意义的束缚。

The iron ore used *to make steel* comes from open-pit and underground mines.
炼钢用的铁矿石来自露天开采的矿山和地下矿井。（不译成"制造钢"）

An insulator offers a very *high resistance* to the passage through which electric current goes.
绝缘体对电流通过有很大阻力。（不译成"高阻力"）

Reading Material

Basics of Safety Engineering

Safety engineering, like any applied science, is based upon fundamental principles and rules of practice. Safety engineering involves the identification, evaluation, and control of hazards in man-machine systems (products, machines, equipment, or facilities) that contain a potential to cause injury to people or damage to property.

1. Hazard Identification

The first step in safety engineering is "hazard identification". A hazard is anything that has the potential to cause harm when combined with some initiating stimulus.

Many system safety techniques have been pioneered to aid in the identification of potential system hazards. None is more basic than "energy analysis". Here, potential hazards associated with various physical systems and their associated operation, including common industrial and consumer products and related activities, can be identified (for later evaluation and control) by first recognizing that system and product "hazards" are directly related to various common forms of "energy". That is, system component or operator "damage" or "injury" cannot occur without the presence of some form of hazardous "energy".

"Hazard identification" in reality can be viewed as "energy identification", recognizing that an unanticipated undesirable release or exchange of energy in a system is absolutely necessary to cause an "accident" and subsequent system damage or operator injury. Therefore, an "accident" can now be seen as "an undesired and unexpected, or at least untimely release, exchange, or action of energy, resulting, or having the potential to result, in system damage or injury". This approach simplifies the task of hazard identification as it allows the identification of hazards by means of a finite set of search paths, recognizing that the common forms of energy that produce the vast majority of accidents can be placed into only ten descriptive categories.

The goal of this first step in the hazard control process is to prepare a list of potential hazards (energies) in the system under study. No attempt is made at this stage to prioritize potential hazards or to determine the degree of danger associated with them. That will come later. At this first stage, one is merely taking an "inventory" of potential hazards (potential hazardous energies). A practical list of hazardous energy types to be identified might include:

(1) **Mechanical Energy Hazards**

Mechanical energy hazards involve system hardware components that cut, crush, bend, shear, pinch, wrap, pull, and puncture. Such hazards are associated with components that move in circular, transverse (single direction), or reciprocating (back and forth) motion. Traditionally, such hazards found in typical industrial machinery have been associated with the terms "power transmission apparatus", "functional components", and the "point of operation".

(2) **Electrical Energy Hazards**

Electrical energy hazards have traditionally been divided into the categories of low voltage electrical hazards (below 440 volts) and high voltage electrical hazards (above 440 volts).

(3) **Chemical Energy Hazards**

Chemical energy hazards involve substances that are corrosive, toxic, flammable, or reactive (involving a release of energy ranging from "not violent" to "explosive" and "capable of detonation").

(4) **Kinetic Energy Hazards**

Kinetic energy hazards involve "things in motion" and "impact", and are associated with the collision of objects in relative motion to each other. This would include impact of objects moving toward each other, impact of a moving object against a stationary object, falling objects, flying objects, and flying particles.

(5) **Potential Energy Hazards**

Potential energy hazards involve "stored energy". This includes things that are under pressure, tension, or compression; or things that attract or repulse one another. Potential energy hazards are associated with things that are "susceptible to sudden unexpected movement". Hazards associated with gravity are included in this category, and involve potential falling objects, potential falls of persons, and the hazards associated with an object's weight. This category also includes the forces transferred to the human body during manual lifting.

(6) **Thermal Energy Hazards**

Thermal energy hazards involve things that are associated with extreme or excessive heat, extreme cold, sources of flame ignition, flame propagation, and heat-related explosions.

(7) **Acoustic Energy Hazards**

Acoustic energy hazards involve excessive noise and vibration.

(8) **Radiant Energy Hazards**

Radiant energy hazards involve the relatively short wavelength energy forms within the electromagnetic spectrum including the potentially harmful characteristics of radar, infra-red, visible, microwave, ultra-violet, x-ray, and ionizing radiation.

(9) **Atmospheric/Geological/Oceanographic Energy Hazards**

These hazards are associated with atmospheric weather circumstances such as wind and storm conditions, geological structure characteristics such as underground pressure or the instability of the earth's surface, and oceanographic currents, wave action, etc.

(10) **Biological Hazards**

These hazards are associated with poisonous plants, dangerous animals, biting insects and disease carrying bacteria, etc.

To develop a list of potential system hazards, one should consider each form of energy in turn. First, list each particular type of energy contained in the system under study, and then describe the various reasonably foreseeable circumstances under which it might become a proximate cause of an undesirable event. Here, full use of the published literature, accident statistics, system operator experience, scientific and engineering probability forecasting, system safety techniques, and team brainstorming are brought to bear on the question of how each form of energy might cause an undesirable event.

Prerequisite to such an identification of all system hazards is a thorough understanding of the system under study related to its general and specific intended purpose and all reasonably anticipated conditions of use.

Specifically, one must thoroughly understand: ①the engineering design of the system, including all physical hardware components—their functions, material properties, operating characteristics, and relationships or interfaces with other system components, ②the intended uses as well as the reasonably anticipated misuses of the system, ③the specific (demographic and human factor) characteristics of intended system users, as well as reasonably anticipated unintended users, taking into account such things as their educational levels, their range of knowledge and skill, and their physical, physiological, psychological, and cultural capabilities, expectancies, and limitations, and ④the general characteristics of the physical and administrative environment in which the system will be operated. That is, one must have a thorough understanding of the man/machine/task/environment elements of the system and their interactions.

2. Hazard Evaluation

The evaluation stage of the safety engineering process has as its goal the prioritizing or ordering of the list of potential system condition or physical state hazards, or potential system personnel of human factors compiled.

The mere presence of a potential hazard tells us nothing about its potential danger. To know the danger related to a particular hazard, one must first examine associated risk factors. Risk can be measured as the product of three components: ①the probability that an injury or damage producing mishap will occur during any one exposure to the hazard; ②the potential severity or degree of injury or damage that will likely result should a mishap occur; and ③the estimated number of times a person or persons will likely be exposed to the hazard over a specific period of time.

In the evaluation of mishap probability, consideration should be given to historical incident data and reasonable methods of prediction.

Use of this equation must take into account that an accident event having a remote probability of occurrence during any single exposure or during any finite period as a result of exposure to a particular hazard is certain to occur if exposure to that hazard is allowed to be repeated over a longer period of time. Therefore, a long-term or large sample view of probability should be taken for proper evaluation.

Determination of severity potential should center on the most likely resulting injury or damage as well as the most severe potential outcome. Severity becomes the controlling factor when severe injury or death is a likely possibility among the several plausible outcomes. That is, even when other risk factors indicate a low probability of mishap over time, if severe injury or death may occur as a result of mishap, the risk associated with such hazards must be considered as being "unacceptable", and strict attention given to the control of such hazards and related mishaps.

Exposure evaluation should consider the typical life expectancy of the system containing a particular hazard, the number of systems in use, and the number of individuals who will be exposed to these systems over time.

This step in the hazard evaluation process will ultimately serve to divide the list of potential hazards into a group of "acceptable" hazards and a group of "unacceptable" hazards. Acceptable hazards are those associated with acceptable risk factors; unacceptable hazards are those associated with unacceptable risk factors.

An "acceptable risk" can be thought of as a risk that a group of rational, well-informed, ethical individuals would deem acceptable to expose themselves to in order to acquire the clear benefits of such exposure. An "unacceptable risk" can be thought of as a risk that a group of rational, well-informed, ethical individuals would deem unacceptable to expose themselves to in order to acquire the exposure benefits.

Hazards associated with an acceptable risk are traditionally called "safe", while hazards associated with an unacceptable risk are traditionally called "unsafe". Therefore, what is called "safe" does contain elements of risk; it is just that such elements have been judged to be "acceptable". Once again, the mere presence of a hazard does not automatically mean that the hazard is associated with any real danger. It must first be measured as being unacceptable.

The result of this evaluation process will be the compiling of a list of hazards (or risks and dangers) that are considered unacceptable. These unacceptable hazards (which render the system within which they exist "unreasonably dangerous") are then carried to the third stage of the safety engineering process, called hazard control.

3. Hazard Control

The primary purpose of engineering and the design of products and facilities is the physical "control" of various materials and processes to produce a specific benefit. The central purpose of safety engineering is the control of system "hazards" which may cause system damage, system user injury, or otherwise decrease system benefits. Current and historic safety engineering references have advocated a specific order or priority in which hazards are best controlled. Listed in order of preference and effectiveness, these control methods have come to be called "cardinal rules of safe design", or the "cardinal rules of hazard control".

The first cardinal rule of hazard control (safe design) is "hazard elimination" or "inherent safety". That is, if practical, one should control (eliminate or minimize) potential hazards by designing them out of products and facilities "on the drawing board". This is accomplished through the use of such interrelated techniques as "hazard removal, hazard substitution, and/or hazard attenuation", through the use of the principles and techniques of system and product safety engineering, system and product safety management, and human factors engineering, beginning with the concept and initial planning stages of the system design process.

The second cardinal rule of hazard control (safe design) is the minimization of system hazards through the use of add-on "safety devices" or "safety features" engineered or designed into products or facilities "on the drawing board" to prevent the exposure of product or facility users to inherent potential hazards or dangerous combinations of hazards; called "extrinsic safety". A sample of such devices would include shields or barriers which guard or enclose hazards, component interlocks, pressure relief valves, stairway handrails, and passive vehicle occupant restraint and crashworthiness systems.

A principle that applies equally to the first two cardinal rules of safe design is that of "passive vs. active" hazard control. Simply, a passive control is a control that works without requiring the continuous or periodic involvement or action of system users. An active control, in contrast, requires the system operator or user to "do something" before system use, continuously or periodically during system operation in order for the control to work and avoid injury. Passive controls are "automatic" controls, whereas active controls can be thought of as "manual" controls. Passive controls are unquestionably more effective than active controls.

The third cardinal rule of hazard control (safe design) is the control of hazards through the development of warnings and instructions; that is, through the development and effective communication of safe system use (and maintenance) methods and procedures that first warn persons of the associated system dangers that may potentially be encountered under reasonably foreseeable conditions of system use, misuse, or service, and then instruct them regarding the precise steps that must be followed to cope with or avoid such dangers.

This third approach must only be used after all reasonably feasible design and safeguarding opportunities (first and second rule applications) have been exhausted.

Further, it must be recognized that the (attempted) control of system hazards through the use of warnings and instructions, the least effective method of hazard control, requires the development of a variety of state-of-the-art communication methods and materials to assure that such warnings and instructions are received and understood by system users.

Among other things, the methods and materials used to communicate required safe use or operating methods and procedures must give adequate attention to the nature and potential severity of the hazards involved, as well as reasonably anticipated user capabilities and limitations (human factors).

Briefly stated, the cardinal rules of hazard control involve system design, the use of physical safeguards, and user training. Further, it must be thoroughly understood that no safety device equals the elimination of a hazard on the drawing board, and no safety procedure equals the use of an effective safety device. This approach has been advocated by the safety literature and successfully practiced by safety professionals for decades.

New Words and Expressions

prioritize [praɪˈɒrɪtaɪz, ˈpraɪərɪ-] v. 把……区分优先次序
pinch [pɪntʃ] n. 捏，撮；收缩；紧急关头
detonation [ˌdetəʊˈneɪʃən] n. 爆炸，爆炸声，爆裂
repulse [rɪˈpʌls] v. 拒绝，排斥；反驳，严拒
propagation [ˌprɒpəˈgeɪʃən] n. 繁殖；传播，宣传
expectancy [ɪkˈspektənsi] n. 期待，期望
rational [ˈræʃənl] adj. 理性的，合理的；推理的
ethical [ˈeθɪkəl] adj. 与伦理有关的；民族的
render [ˈrendə] v. 呈递；归还；着色
cardinal [ˈkɑːdɪnəl] adj. 主要的，最重要的
attenuation [əˌtenjʊˈeɪʃən] n. 变薄，变细；衰减
extrinsic [eksˈtrɪnsɪk] adj. 外在的，外表的，外来的

Unit Three

The Ergonomics Process

1. Description of the Ergonomics Process

The name "ergonomics" comes from the Greek words "ergon", which means work and "nomos" which means law. Ergonomics is the study of the interaction between people and machines and the factors that affect the interaction. Its purpose is to improve the performance of systems by improving human-machine interaction. This can be done by "designing-in" a better interface or by "designing-out" factors in the work environment, in the task or in the organization of work that degrade human-machine performance.

Systems can be improved by:
- Designing the user-interface to make it more compatible with the task and the user. This makes it easier to use and more resistant to errors that people are known to make.
- Changing the work environment to make it safer and more appropriate for the task.
- Changing the task to make it more compatible with user characteristics.
- Changing the way work is organized to accommodate people's psychological, and social needs.

In an information processing task, we might redesign the interface so as to reduce the load on the user's memory (e. g. shift more of the memory load of the task onto the computer system or redesign the information to make it more distinctive and easier to recall). In a manual handling task, we might redesign the interface by adding handles or using lighter or smaller container to reduce the load on the musculoskeletal system. Work environments can be improved by eliminating vibration and noise and providing better seating, desking, ventilation or lighting, for example. New tasks can he made easier to learn and to perform by designing them so that they resemble tasks or procedures that people are already familiar with. Work organization can be improved by enabling workers to work at their own pace, so as to reduce the psychophysical stresses of being "tied to the machine" or by introducing subsidiary tasks to increase the range of physical activity at work and provide contact with others.

The implementation of ergonomics in system design should make the system work better by eliminating aspects of system functioning that are undesirable, uncontrolled or unaccounted for, such as:

- Inefficiency—when worker effort produces sub-optimal output.
- Fatigue—in badly designed jobs people tire unnecessarily.
- Accidents, injuries and errors—due to badly designed interfaces and/or excess stress either mental or physical.
- User difficulties—due to inappropriate combinations of subtasks making the dialogue/interaction cumbersome and unnatural.
- Low morale and apathy.

In ergonomics, absenteeism, injury, poor quality and unacceptably high levels of human error are seen as system problems rather than "people" problems, and their solution is seen to lie in designing a better system of work rather than in better "man management" or incentives, by "motivating" workers or by introducing safety slogans and other propaganda.

2. The Focus of Ergonomics

The focus is on the interaction between the person and the machine and the design of the interface between the two. Every time we use a tool or a machine we interact with it via an interface (a handle, a steering wheel, a computer keyboard and mouse, etc.). We get feedback via an interface (the dashboard instrumentation in a car, the computer screen, etc.) The way this interface is designed dertermines how easily and safely we can use the machine.

When faced with productivity problems, engineers might call for better machines, personnel management might call for better-trained people. Ergonomists call for a better interface and better interaction between the user and the machine-better task design.

3. Human-Machine Systems

A system is a set of elements, the relations between these elements and the boundary around them. Most systems consist of people and machines and perform a function to produce some form of output. Inputs are received in the form of matter, energy and information. For ergonomics, the human is part of the system and must be fully integrated into it at the design stage. Human requirements are therefore system requirements, rather than secondary considerations and can be stated in general terms as requirements for:
- Equipment that is usable and safe.
- Tasks that are compatible with people's expectations, limitations and training.
- An environment that is comfortable and appropriate for the task.
- A system of work organization that recognizes people's social and economic needs.

4. Compatibility-Matching Demands to Capabilities

Compatibility between the user and the rest of the system can be achieved at a number of levels. Throughout this book we will encounter compatibility at the biomechanical, anatomical, physiological, behavioral and cognitive levels. It is a concept that is common to the application of ergonomics across a wide range of settings and disciplines. In order to achieve compatibility, we need to assess the demands placed by the technological and environmental constraints and weigh them against the capabilities of the users. The database of modern ergonomics contains much information on the capabilities and characteristics of people and one of the main purposes of this book is to introduce the reader to this information and show how it can be used in practice.

Ergonomic entropy (Karwowski et al., 1994) is disorder in system functioning that occurs owing to a lack of compatibility in some or all of the interactions involving the human operator. This incompatibility can occur for a variety of reasons, for example:

- Human requirements for optimum system functioning were never considered at the design stage (e.g. There was a failure to consult appropriate standards, guide-lines or textbooks).
- Inappropriate task design (e.g. New devices introduce unexpected changes in the way tasks are carried out and these are incompatible with user knowledge, habits or capacity, or they are incompatible with other tasks).
- Lack of prototyping (e.g. Modern software development is successful because it is highly iterative; users are consulted from the conceptual stage right through to pre-production prototypes).

5. Application of Ergonomics

The purpose of ergonomics is to enable a work system to function better by improving the interactions between users and machines. Better functioning can be defined more closely, for example, as more output from fewer inputs to the system (greater "productivity") or increased reliability and efficiency (a lower probability of inappropriate interactions between the system components). The precise definition of better functioning depends on the context. Whatever definition is used should, however, be made at the level of the total work system and not just one of the components.

Improved machine performance that increased the psychological or physical stress on workers or damaged the local environment would not constitute improved performance of the total work system or better attainment of its goals. Workstation redesign to make workers more "comfortable" is an incorrect reason for the application of ergonomics if it is done superficially, for its own sake, and not to improve some aspect of the functioning of the total work system (such as reduced absenteeism and fewer accidents due to better working conditions).

There are two ways in which ergonomics impacts upon systems design in practice. Firstly, many ergonomists work in research organizations or universities and carry out basic research to discover the characteristics of people that need to be allowed for in design. This research often leads, directly or indirectly, to the drafting of standards, legislation and design guidelines. Secondly, many ergonomists work in a consultancy capacity either privately or in an organization. They work as part of a design team and contribute their knowledge to the design of the human-machine interactions in work systems. This often involves the application of standards guidelines and knowledge to specify particular characteristics of the system. Real work systems are hierarchical. This means that the main task is made up of sub-tasks (the next level down) and is governed by higher-level constraints that manifest as style of supervision, type of work organization, working hours and shiftwork, etc. If we want to optimize a task in practice, we rarely redesign the task itself. We either change or reorganize the elements of the task (at the next level down) or we change the higher-level variables. For example, to optimize a data entry task we might look at the style of human computer dialogue that has been chosen. We might find that there are aspects of the dialogue that cause errors to be transmitted to the system

(e. g. when the operator mistakenly reverses two numbers in a code, the system recognizes it as a different code rather than rejecting it). Alternatively, we might find there are insufficient rest periods or that most errors occur during the night shift.

To optimize a task we first have to identify the level of the task itself (e. g. a repetitive manual handling task), the next level down (the weight and characteristics of the load and the container) and the next level up (the workload and work organization). To optimize the task we can either redesign it from the bottom up (e. g. use lighter containers and stabilize the load) or from the top down (e. g. introduce job rotation or more rest periods) or both. At the same time we can look at extraneous or environmental factors at the level of the task but external to it, factor, that also degrade performance (e. g. slippery floors in the lifting example or bad lighting and stuffy air in the data entry example).

Having redesigned the task and evaluated the improvements to task performance, we then monitor it over time to detect improvements in system performance.

One of the problems facing the ergonomist both in the design of new work systems and in the evaluation of existing ones is to ensure that all aspects are considered in a systematic way. The human-machine approach enables key areas to be identified irrespective of the particular system so that ergonomics can be applied consistently in different systems.

The first step is to describe the work system and its boundaries. This enables the content and scope of the application of ergonomics to be specified. Next, the human and machine components and the local environment are defined and described in terms of their main components. Following this, the interactions between the various components can be analyzed to identify the points of application of basic knowledge to the design/evaluation process. Examples of interactions are the interaction between the displays and the workspace—this directs attention to the positioning of the displays in the workspace so that the operator can see them when working. The interaction between the effectors and the workspace introduces considerations about the space requirements for body movements required by the task.

One of the problems facing the ergonomist both in the design of new work systems and in the evaluation of existing ones is to ensure that all aspects are considered in a systematic way. The human-machine approach enables key areas to be identified irrespective of the particular system so that ergonomics can be applied consistently in different systems.

New Words and Expressions

musculoskeletal [ˌmʌskjʊləʊˈskelɪt(ə)l]   adj. 肌(与)骨骼的
ventilation [ˌventɪˈleɪʃ(ə)n]   n. 通风，通风设备
subsidiary [səbˈsɪdɪəri]   adj. 附属的，辅助的
apathy [ˈæpəθi]   n. 冷漠无情，漠不关心
absenteeism [ˌæbs(ə)nˈtiːɪz(ə)m]   n. 有计划的怠工，经常无故缺席
propaganda [ˌprɒpəˈɡændə]   n. 宣传
dashboard [ˈdæʃbɔːd]   n. 仪表板，仪表盘
optimum [ˈɒptɪməm]   adj. 最佳状态的，最适宜的
prototype [ˈprəʊtətaɪp]   n. 技术原型，技术雏形
hierarchical [haɪəˈrɑːkɪk(ə)l]   n. 分层的，分等级的；具有层级关系的
irrespective [ˌɪrɪˈspektɪv]   adj. 不谈论的，不考虑的

Translation Skill

科技英语翻译技巧（二）——词量增减

在英译汉时，根据汉语的习惯，在译文中可增加一些原文中无其形而有其意的词，或减去原文中某些在译文中属于多余的词。

一、词量增加

Matter can be changed into energy, and energy into matter.
物质可以转化为能，能也可以转化为物质。（增补英语中省略的词）

The best conductor has the least resistance and the poorest has the greatest.
最好的导体电阻最小，最差的导体电阻最大。（增补英语中省略的词）

If A is equal to D, A plus B equals D plus B.
若 A = D，则 A + B = D + B（增加关联词）。

The first electronic computers used vacuum tubes and other components, and this made the equipment very large and bulky.
第一代电子计算机使用真空管和其他元件，这使得设备又大又笨。（增加表示复数的词）

Oxidation will make iron and steel rusty.
氧化作用会使钢铁生锈。（增加具有动作意义的抽象名词）

The cost of such a power plant is a relatively small portion of the total cost of the development.
这样一个发电站的修建费用仅占该开发工程总费用的一小部分。（增加具有动作意义的抽象名词）

The resistance of the pipe to the flow of water through it depends upon the length of the pipe, the diameter of the pipe, and the feature of the inside walls (rough or smooth).
水管对通过的水流的阻力取决于下列三个因素：管道长度、管道直径、管道内壁的特性（粗糙或光滑）。（补充概括性的词）

Heat from the sun stirs up the atmosphere, generating winds.
太阳发出的热能搅动大气，于是产生了风。（修辞加词，语气连贯）

In general, all the metals are good conductors, with silver the best and copper the second.
一般来说，金属都是良导体，其中以银为最好，铜次之。（修辞加词，语气连贯）

二、词量减少

在英译汉时，根据汉语的习惯可将冠词、介词、连词、动词、名词、人称代词、物主代词、反身代词、关系代词等省译。

The *world* of work injury insurance is complex.
工伤保险是复杂的。（名词省译）

Any substance is made of atoms whether it is *a* solid, *a* liquid, or *a* gas.
任何物质，不论是固体、液体或气体，都是由原子组成的。（冠词省译）

In the absence *of* force, a body will either remain *at* rest, or continue to move *with* constant speed *in* a straight line.
无外力作用，物体则保持静止状态，或做匀速直线运动。（介词省译）

A wire lengthens while *it* is heated.

金属丝受热则伸长。(代词省译)
Practically all substances expand when heated *and* contract when cooled.
几乎所有的物质都是热胀冷缩的。(连词省译)
Stainless steels *possess* good hardness and high strength.
不锈钢硬度大、强度高。(动词省译)

Reading Material

Implementation of Human Error Diagnosis System

1. Introduction

As the manufacturing strategies progress toward collaborative manufacturing partnerships, the relationship between suppliers and customers forms a tightly coupled business organization. Any unexpected event, such as industrial accident, can initiate a chain reaction, which might endanger many enterprise survivals. Human errors, defined as the unsafe acts performed by operators or decision makers, have been generally recognized as the major cause for industrial accidents. The impacts of these accidents depend on the magnitude of deviation from normal situations and the robustness of system. The severity of human error problem is worsen by the increasing mental workload due to the changing role of human in modern manufacturing environment, and the insufficient company resources spend due to the lack of understanding of the human error cost. Therefore, the human error control techniques gradually become important tools for industrial managers to ensure the agile manufacturing activities. To prevent the accidents from breaking down the tightly coupled manufacturing activities, it is necessary for managers to realize the potential human error threats from daily operations. It is necessary to provide factory managers with tools to identify human errors and estimate related cost. These tools aim to control the fluctuations of manufacturing activities caused by unstable human performances within an acceptable risk level. Traditionally, task analysis techniques were adopted to explore the human-machine system faults, which might create problems for human to make mistakes. Job analysis techniques were improvised to set the criteria for selecting the right person for the job. These techniques require the collection of system information and standard operation procedures and the assistance from ergonomists or safety experts. Once the targeted jobs were analyzed, the potential human errors are identified and controlled through the implementation of appropriate preventive measurements.

Unfortunately, traditional approach only accesses the static safety status of the factory. Factory facilitates and operations will not remain the same condition as they were assessed. Facilities may deteriorate because of the natural wear-off, and the operation conditions may fluctuate because of the human inconsistence. Consequently, the system safety status is not in a static but a dynamic condition. Moreover, not every company can afford the expense of hiring its own experts. Not to mention the qualified personnel are sometimes not easy to acquire. As a result, companies can realize the human error problems in their working environment only when things went wrong. Factory operated under an acceptable level of risk is still in the cloud of unacceptable accidents, which, if happen, will disrupt the tightly coupled production activities. This study proposed a systematical approach to implement human error diagnosis (HED) system to help industry to develop its own human error assessment model based on the evaluation of daily plant operations. This system provides the decision makers with the capability to assess the human errors and their impacts in manufacturing environment so that accidents can be avoided.

2. Researches Related to Human Error Identification Techniques

To prevent the accidents from happening, it is necessary for factory supervisors to access the risk caused by human errors from daily operations and implement the appropriate measurements. This means the supervisors must have tools to identify the potential human errors and assess their impacts. Various human error data banks have been developed, using either empirical or human estimated approach, to predict the human errors. None of them are comprehensive enough to cover industrial cases. Some researchers proposed to develop generic tools to identify human errors. Kirwan classified these human error identification techniques into five categories: taxonomies methods, psychologically based tools, cognitive modeling tools, cognitive simulation tools, and reliability-oriented tools. Kirwan argued that there are currently no totally comprehensive human error analysis (HEA) tools. Most tools can only be used as a supplementary tool for human error analyzer. The achievements of a comprehensive error analysis depend on the assessor's own judgment. The experiences of human error analyzer become an important factor for the successful identification process. Strater pointed out that when performing human error identification, in order to fulfill the generic requirements of data, it is important to collect empirical plant-specific data. The above researches indicate that the plant experience about human failure and human performance should be used to support the process of analyzing and assessing human reliability.

In practice, the factory managers join the risk assessment projects and provide the current facility and operation data to the fellow assessment teams. This approach leads to the unsatisfied results that the managers have few knowledge of the system daily safety operation status. T. H. Liu and S. L. Hwang developed a systematic approach, called HED system, to help the industry develop its own human error diagnosis system, which can help factory managers to identify the human errors specially existing in their daily operations. The HED system adopted the human error identification in systems tool (HEIST) to identify human errors. Through a systematic process performed by a group of plant managers, human error checklists are generated as data collection tools. Several methods have been proposed to identify the critical human tasks, such as the criticality model, the risk priority number (RPN) model, and the expected cost (EC) model. F. J. Yu and S. L. Hwang pointed out that these approaches might have limitations in finding the human error modes through evaluations of operation procedures. They developed human error cost analysis (HECA) method to identify critical task and error modes through the evaluation of standard of assembly procedure (SAP). This approach uses human error information derived by a group of senior workers and managers based on their experiences and judgments as the source of analysis. However, the daily operation status was not included in the process. In the present study, the error data are collected through observing the workers' operation status, instead of experts' judgments, and can be used as the source information for analyzing human error and developing safety index. Safety index is used as a monitor tool to control human errors. In this way, the operators' behaviors can be incorporated into the human error assessment process and the factory managers can get closer understanding of their manufacturing risk.

3. A Theoretical Framework of Implementing HED System

The implementation of the human error identification system is divided into two phases: I) human error data collection and II) safety index analysis. The main task of phase I is to develop human error checklists through the application of HED method. This includes two stages: preprocessing and identification. The tasks performed in the preprocessing stage are: selecting a target operation, collecting the standard of target operation, verifying the standard of target operation with the real working procedure, and adjusting the operation procedure. In identification stage, potential human error are identified through the standard operation procedure by using the HEIST. HEIST is a theoretical approach, which is, argued by Kirwan, a potential extent of a comprehensive and functional error reduction-based human error analysis approach. This approach aims to identify the external error mode through "free-form" identification or via a set of question-answer routines. The HED developers consider each table for all tasks involved. The analyzer takes the operation procedure as input data and generates the potential human errors tables. The error mode tables are used to develop the human error diagnosis checklists that can be conducted by the managers to observe workers' daily operations.

The main task performed in phase II is to calculate the safety index. Safety index is defined as the multiplication of human error probability and human error effect probability. First, the collected operators' data are converted into human error probability (HEP) by dividing the numbers of nonconformity with the total number of observations. The corresponding human error modes and their error effects are derived from previous HEIST through the help of experts' judgments. The error effect probability table adapted from MIL-STD-1629A is used to convert the four qualitative levels judged by expert group with quantitative values associated with each of these levels.

4. The Implementation of the HED System

Human Error Data Collection. This study was conducted in a semiconductor plant. Since 1996, the occupational incidents have cost the semiconductor industry more than 20 billions of losses. The facility studied in this research was gas cabinet. Gas cabinet is a metal enclosure that is intended to provide local exhaust ventilation for protection the gas cylinder from fire and for protection the surroundings from fire within. The silane gas cabinet is chosen as the study object because it has been widely installed in semiconductor plants and several accidents involved in the past. Case histories involving silane incidents in the semiconductor industry include: property loss, contamination, operator injuries, explosions, and death. The standard operation procedures (SOP) of silane gas cabinet vary by equipment discrepancies.

This study investigates the 3-CYL Cabinet manufactured and assembled by air liquid electronic (ALE). The implementation of the HED system began with phase I: preprocessing stage. The SOP was documented based on the on-site operation manual, observation, and interview results. The observation subject is a male supervisor with five years of related experiences. The SOP can be divided into four stages: pre-purge, cylinder change, post purge, and process flow stages. Pre-purge stage consists of 10 steps. Cylinder change stage consists of 13 steps. Post purge stage consists of 3 steps. Process flow stage consists of 8 steps. Totally 34 steps are classified into four principle stages by using the HEIST based on a group of six analyzers' judgments. The members of the analyzer include one ergonomist, two safety engineers, one facility engineer and two maintenance engineers. All of them are currently working in semiconductor related industries. This group first accepts the HEIST training to be familiar with the analysis. Then the accident cases of silane gas cabinet, SOP and P&ID were studied. The group identified external error modes for each error-identifier prompt generated from previously selected stages by considering the potential errors when operating the facility. Repeating the above process until all 34 steps were analyzed generated sets of human error tables. The human error tables were verified and revised by interviewing two senior gas room supervisors both with more than five years of silane gas cabinet operation experiences.

The first step in the pre-purge stage is to make sure the cylinder in the gas supply side is in the "Process Gas Flow" state and the supply pressure is in "Normal" condition. This step is classified as "Identification of System State". The possible system cause/psychological error-mechanism for identification of step 1 are failing to consider special circumstances, slips of memory or inadequate mental model. The error reduction guidelines provided by HEIST include training procedure, special circumstances, implement shift-technical-advisor role, and local warnings in the interface displays/controls. In this case, the training course and a procedure checklist are provided to reduce this type of error. Continue analyzing the rest of operation steps until all 34 steps are investigated. Totally 174 different types of error modes are identified. These human error tables are converted into human sample checklist.

Safety Index Analysis. The frequency of changing silane cylinder varies due to the different material consumption rate of the production line. To avoid the interruption of production activities, each gas cabinet is equipped with reserved cylinder and automatic switching device, which will shutdown the empty cylinder and turn on the reserved cylinder. Automatic alarm system will signal the gas room supervisor to change the empty cylinder. Then gas suppliers were informed to install the new cylinder within 24 hours. Every time the new cylinder was installed on the gas cabinet, safety procedures for testing the leakage must be performed. During the preliminary sampling period, the frequency of performing the silane cylinder change is one cabinet per day.

The changing operation requires two persons to complete the task. One person performs the changing activities and the other one monitors the operation for safety precaution. In this study, the analyzer observed the operation without intervening the workers' operation. Workers were explained the purpose of this study before the observation to avoid the change of working behavior due to the worry of being evaluated. Line workers also verified the checklists to increase their understanding of this study. The analyzer observed the worker's operation step by step and judged the worker's performance as either conforming or nonconforming on the basis of whether or not they violate the potential human errors.

Observer based on the evaluation results of real situation also estimated the corresponding error effect probability. One safety engineering and one ergonomist, both with at least three years of gas room operation management experiences, were recruited to estimate the human error effect probability. The error effects were classified into four different levels: actual loss, probable loss, possible loss and no effect. The respective qualitative levels of error effect were assigned to each error mode based on these two experts' judgments. Then, the quantitative values were converted by using error effect table. There are 25 silane gas cabinet cylinder change operations during one-month data collection period. The human error probability for each error mode is total number of violations divided by total number of observations. The safety index for each operation step is the summation of corresponding HEP multiplied by error effect probability for respective error mode.

New Words and Expressions

agile ['ædʒaɪl]	adj. 敏捷的，轻快的，灵活的
deteriorate [dɪ'tɪərɪəreɪt]	v. （使）恶化
empirical [em'pɪrɪkəl]	adj. 完全根据经验的
MIL-STD	n. （美军）军用标准；MIL = Military Specifications 军用规格；STD = Standard 标准
cylinder ['sɪlɪndə]	n. 圆柱体；气缸
the 3-CYL Cabinet	n. 一种（有抽屉或格子的）通风橱柜
purge [pɜːdʒ]	n. 净化；清除
reserved [rɪ'zɜːvd]	adj. 保留的，预订的
precaution [prɪ'kɔːʃən]	n. 预防，警惕，防范
recruit [rɪ'kruːt]	v. 使恢复；补充
summation [sʌ'meɪʃən]	n. [数] 总和，和，合计

Unit four

Hazard Identification

Hazard identification is a process controlled by management. You must assess the outcome of the hazard identification process and determine if immediate action is necessary or if, in fact, there is an actual hazard involved. When you do not view a reported hazard as an actual hazard, it is critical to the ongoing process to inform the worker that you do not view it as a true hazard and explain why. This will insure the continued cooperation of workers in hazard identification.

It is important to remember that a worker may perceive something as a hazard, when in fact it may not be a true hazard; the risk may not match the ranking that the worker placed on it. Also, even if a hazard exists, you need to prioritize it according to the ones that can be handled quickly, which may take time, or which will cost money above your budget. If the correction will cause a large capital expense and the risk is real but does not exhibit an extreme danger to life and health, you might need to wait until next year's budget cycle. An example of this would be when workers complain of a smell and dust created by a chemical process. If the dust is not above accepted exposure limits and the smell is not overwhelming, then the company may elect to install a new ventilation system, but not until the next year because of budgetary constraints. The use of PPE until hazards can be removed may be required.

The expected benefits of hazard identification are a decrease in the incidents of injuries, a decrease in lost workdays and absenteeism, a decrease in workers' compensation costs, increased productivity, and better cooperation and communication. The baseline for determining the benefit of the hazard identification can be formulated from existing company data on occupational injuries/illnesses, workers' compensation, attendance, profit, and production.

Hazard identification includes those items that can assist you with identifying workplace hazards and determining what corrective action is necessary to control them. These items include jobsite safety inspections, accident investigations, safety and health committees, and project safety meetings. Identification and control of hazards should include periodic site safety inspection programs that involve supervisors and, if you have them, joint labor management committees. Safety inspections should ensure that preventive controls are in place (PPE, guards, maintenance, engineering controls), that action is taken to quickly address hazards, that technical resources such as Occupational Safety and Health Administration (OSHA), state agencies, professional organizations, and consultants are used, and that safety and health rules are enforced.

Many workplaces have high accident incidence and severity rates because they are hazardous. Hazards are dangerous situations or conditions that can lead to accidents. The more hazards present, the greater the chance that there will be accidents. Unless safety procedures are followed, there will be a direct relationship between the number of hazards in the workplace and the number of accidents that will occur there.

As in most industries, people work together with machines in an environment that causes employees to face hazards, which can lead to injury, disability, or even death. To prevent industrial accidents, the people, machines, and other factors which can cause accidents, including the energies associated with them, must be controlled. This can be done through education and training, good safety engineering, and enforcement.

The core of an effective safety and health program is hazard identification and control. Periodic inspections and procedures for correction and control provide methods of identifying existing or potential hazards in the workplace and eliminating or controlling them. The hazard control system provides a basis for developing safe work procedures and injury and illness prevention training. Hazards occurring or recurring reflect a breakdown in the hazard control system.

The written safety and health program establishes procedures and responsibilities for the identification and correction of workplace hazards. The following activities can be used to identify and control workplace hazards: hazard reporting system, job site inspections, accident investigation, and expert audits.

After all basic steps of the operation of a piece of equipment or job procedure have been listed, we need to examine each job step to identify hazards associated with each job step. The purpose is to identify and list the possible hazards in each step of the job. Some hazards are more likely to occur than others, and some are more likely to produce serious injuries than others. Consider all reasonable possibilities when identifying hazards.

1. Accident Types

(1) **Struck-Against Type of Accidents**

Look at the first four basic accident types—struck-against, struck-by, contact-with and contacted-by—in more detail, with the job step walk-round inspection in mind. Can the worker strike against anything while doing the job step? Think of the worker moving and contacting something forcefully and unexpectedly—an object capable of causing injury. Can he or she forcefully contact anything that will cause injury? This forceful contact may be with machinery, timber or bolts, protruding objects or sharp, jagged edges. Identify not only what the worker can strike against, but how the contact can come about. This does not mean that every object around the worker must be listed.

(2) **Struck-By Type of Accidents**

Can the worker be struck by anything while doing the job step? The phrase "struck by" means that something moves and strikes the worker abruptly with force. Study the work environment for what is moving in the vicinity of the worker, what is about to move, or what will move as a result of what the worker does. Is unexpected movement possible from normally stationary objects? Examples are ladders, tools, containers, and supplies.

(3) Contact-By and Contact-With Types of Accidents

The subtle difference between contact-with and contact-by injuries is that in the first, the agent moves to the victim, while in the second, the victim moves to the agent.

Can the worker be contacted by anything while doing the job step? The contact-by accident is one in which the worker could be contacted by some object or agent. This object or agent is capable of injuring by nonforceful contact. Examples of items capable of causing injury are chemicals, hot solutions, fire, electrical flashes, and steam.

Can the worker come in contact with some agent that will injure without forceful contact? Any type of work that involves materials or equipment that may be harmful without forceful contact is a source of contact-with accidents. There are two kinds of work situations which account for most of the contact-with accidents. One situation is working on or near electrically charged equipment, and the other is working with chemicals or handling chemical containers.

(4) Caught-In and Caught-On Types of Accidents

The next three accident types involve "caught" accidents. Can the person be caught in, caught on, or caught between objects? A caught-in accident is one in which the person, or some part of his or her body, is caught in an enclosure or opening of some kind. Can the worker be caught on anything while doing the job step? Most caught on accidents involve worker's clothing being caught on some projection of a moving object. This moving object pulls the worker into an injury contact. Or, the worker may be caught on a stationary protruding object, causing a fall.

(5) Caught-Between Type of Accidents

Can the worker be caught between any objects while doing the job step? Caught-between accidents involve having a part of the body caught between something moving and something stationary, or between two moving objects. Always look for pinch points.

(6) Fall-to-Same-Level and Fall-to-Below Types of Accidents

Slip, trip, and fall accident types are some of the most common accidents occurring in the workplace. Can the worker fall while doing a job step? Falls are such frequent accidents that we need to look thoroughly for slip, trip, and fall hazards. Consider whether the worker can fall from something above ground level, or whether the worker can fall to the same level.

Two hazards account for most fall-to-same level accidents: slipping hazards and tripping hazards. The fall-to-below accidents occur in situations where employees work above ground or above floor level, and the results are usually more severe.

(7) Overexertion and Exposure Types of Accidents

The next two accident types are overexertion and exposure. Can the worker be injured by overexertion; that is, can he or she be injured while lifting, pulling, or pushing? Can awkward body positioning while doing a job step cause a sprain or strain? Can the repetitive nature of a task cause injury to the body? An example of this is excessive flexing of the wrist, which can cause carpal tunnel syndrome (which is abnormal pressure on the tendons and nerves in the wrist).

Finally, can exposure to the work environment cause injury to the worker? Environmental conditions such as noise, extreme temperatures, poor air, toxic gases and chemicals, or harmful fumes from work operations should also be listed as hazards.

2. Hazard Reporting System

Hazard identification is a technique used to examine the workplace for hazards with the potential to cause accidents. Hazard identification, as envisioned in this section, is a worker-oriented process. The workers are trained in hazard identification and asked to recognize and report hazards for evaluation and assessment. Management is not as close to the actual work being performed as are those performing the work. Even supervisors can use extra pairs of eyes looking for areas of concern.

Workers have already hazard concerns and have often devised ways to mitigate the hazards, thus preventing injuries and accidents. This type of information is invaluable when removing and reducing workplace hazards.

This approach to hazard identification does not require that someone with special training conduct it. It can usually be accomplished by the use of a short fill-in-the-blank questionnaire. This hazard identification technique works well where management is open and genuinely concerned about the safety and health of its workforce. The most time-consuming portion of this process is analyzing the assessment and response regarding potential hazards identified. Empowering workers to identify hazards, make recommendations on abatement of the hazards, and then suggest how management can respond to these potential hazards is essential.

New Words and Expressions

perceive [pəˈsiːv]	v.	察觉；感知，感到，认识到
overwhelming [ˌəʊvəˈwelmɪŋ]	adj.	压倒性的；无法抵抗的
budgetary [ˈbʌdʒɪtəri]	adj.	预算的
constraint [kənˈstreɪnt]	n.	约束
audit [ˈɔːdɪt]	n.	审计，核查
bolt [bəʊlt]	n.	螺钉
protrude [prəˈtruːd]	v.	突出
jagged edge		锯齿边缘
in the vicinity of		在邻近
subtle [ˈsʌtl]	adj.	微妙的；精细的
stationary [ˈsteɪʃ(ə)nəri]	adj.	固定的
pinch [pɪntʃ]	n.	收缩；压力
trip [trɪp]	n.	绊倒，摔倒，失足
awkward [ˈɔːkwəd]	adj.	难使用的；笨拙的
sprain [spreɪn]	n.	扭伤，扭筋
strain [streɪn]	n.	过劳损伤
flexing [ˈfleksɪŋ]	n.	挠曲，可挠性
wrist [rɪst]	n.	手腕，腕关节
carpal [kɑːpl]	n.	腕骨
syndrome [ˈsɪndrəʊm]	n.	综合病症
tendon [ˈtendən]	n.	[解] 腱
nerve [nɜːv]	n.	神经
fume [fjuːm]	n.	（浓烈或难闻的）烟，气体

envision [ɪnˈvɪʒən]　　　　　　　　v. 想象，预想
questionnaire [ˌkwestʃəˈneə(r)]　　n. 调查表，问卷
genuinely [ˈdʒenjuɪnli]　　　　　　adv. 真诚地，诚实地
time-consuming [ˈtaɪmkənˌsjuːmɪŋ]　adj. 耗时的
empower [ɪmˈpauə]　　　　　　　v. 授权给，使能够

Translation Skill

科技英语翻译技巧（三）——词性转换

因为英汉两种语言属于不同的语系，所以它们在语言结构与表达形式方面各有特点。要使译文既忠实于原意，又顺畅可读，就不能局限于逐词对等，必须采用适当的词性转换。

一、转译成汉语动词

A change of state from a solid to a liquid form requires heat energy.
从固态变为液态需要热能。（名词转译成动词）
The term laser stands for amplification by stimulated emission of radiation.
"激光"这个术语指的是利用辐射的受激发射放大光波。（介词转译成动词）
Both of the substances are not soluble in water.
这两种物质都不溶于水。（形容词转译成动词）
In this case the temperature in the furnace is up.
在这种情况下，炉温就升高。（形容词转译成动词）

二、转译成汉语名词

Such materials are characterized by good insulation and high resistance to wear.
这些材料的特点是：绝缘性好，耐磨性强。（动词转译成名词）
The result of this experiment is much better than those of previous ones.
这次实验的结果比前几次的实验结果好得多。（代词转译成名词）
All structural materials behave plastically above their elastic range.
超过弹性极限时，一切结构材料都会显示出塑性。（副词转译成名词）
某些表示事物特征的形容词做表语时可将其转译成名词，其后往往加上"性""度""体"等。带有定冠词的某些形容词用作名词时，应译成名词。
The cutting tool must be strong, tough, hard, and wear resistant.
刀具必须具有足够的强度、硬度、韧性和耐磨性。
Both the compounds are acids, the former is strong, the latter is weak.
这两种化合物都是酸，前者是强酸，后者是弱酸。

三、转译成汉语形容词

This man-machine system is chiefly characterized by its simplicity of operation and the ease with which it can be maintained.
这种人机系统的主要特点是操作简单，容易维修。（副词转译成形容词）
It is demonstrated that dust is extremely hazardous.
已经证实，粉尘具有极大的危害。（副词转译成形容词）

The equations below are derived from those *above*.
下面的这些方程式是由上面的那些方程式推导出来的。（副词转译成形容词）
This experiment was a *success*.
这个试验是成功的。（名词转译成形容词）
四、转译成汉语副词
The mechanical automatization makes for a *tremendous* rise in labor productivity.
机械自动化可以大大地提高劳动生产率。（形容词转译成副词）
A helicopter is *free* to go almost anywhere.
直升机几乎可以自由地飞到任何地方去。（形容词转译成副词）
Rapid evaporation at the heating surface *tends* to make the steam wet.
加热面上的迅速蒸发，往往使蒸汽的湿度变大。（动词转译成副词）

Analyzing Hazards

1. Hazard Analysis

Hazard analysis is a technique used to examine the workplace for hazards with the potential to cause accidents. The information obtained by the hazard identification process provides the foundation for making decisions upon which jobs should be altered in order for die worker to perform the work safely and expeditiously. Also, this process allows workers to become more involved in their own destiny. For some time, involvement has been recognized as a key motivator of people. This is also a positive mechanism in fostering labor/management cooperation. This is especially true if everyone in the workplace is continuously looking for the potential hazards which can result in injury, illness or even death.

Hazard analysis can get pretty sophisticated and go into much detail. Where the potential hazards are significant and the possibility for trouble is quite real, such detail may well be essential. However, for many processes and operations—both real and proposed—a solid look at the operation or plans by a variety of affected people may be sufficient.

Analysis often implies mathematics, but calculating math equations is not the major emphasis when attempting to address hazards or accidents/incidents which occur within the industry analysis in the context of this module means taking time to examine systematically the worksite's existing or potential hazards. This can be accomplished in a variety of ways.

If you are faced with fairly sophisticated and complex risks with a reasonable probability of disaster if things go wrong, you may want some help with some of the other hazard analysis methodologies. What follows is a very brief look at the common ones. If you decide to try one of the approaches, check with your local OSHA consultation office or call an engineering firm which specializes in hazards analysis.

2. Root Cause Analysis

Accidents are rarely simple and almost never result from a single cause. Accidents may develop from a sequence of events involving performance errors, changes in procedures, oversights, and omissions. Events and conditions must be identified and examined in order to find the cause of the accident and a way to prevent that accident and similar accidents from occurring again. To prevent the recurrence of accidents one must identify the accident's causal factors. The higher the level in the management and oversight chain in which the root cause is found, the more diffused the problem can be.

Root cause analysis aids in the development of evidence, by collecting information and putting the information in the logical sequence so that it can be easily examined. This will lead to the causal factors of the accident and then to a development of new methods in order to help eliminate that accident or similar accidents from recurring in the future. By creating an event in the causal factor chain, multiple causes can be visually illustrated and a visual relationship between the direct and contributing causes can be shown. Event causal charting also visually delineates the interactions and relationships of all involved groups and/or individuals. By using root cause analysis, one can develop an event causal chain to examine the accident in a step-by-step manner by looking at the events, conditions, and causal factors chronologically, in order to prevent future accidents.

Root cause analysis is used when there are multiple problems with a number of causes of an accident. A root cause analysis is a sequence of events that shows, step-by-step, the events that took place in order for the accident to occur. Root cause analysis puts all the necessary and sufficient events and causal factors for an accident in a logical, chronological sequence. It analyzes the accident and evaluates evidence during an investigation. It is also used to help prevent similar accidents in the future and to validate the accuracy of preaccidental system analysis. It is used to help identify an accident's causal factors which, once identified, can be fixed to eliminate future accidents of the same or similar nature.

On the downside, root cause analysis is time-consuming and requires the investigator to be familiar with the process for it to be effective. As you will see later in this unit, you may need to revisit an accident scene multiple times and also look at areas that are not directly related to the accident in order to have a complete event and causal factor chain. Analysis requires a broad perspective of the accident in order to identify any hidden problems that would have caused the accident.

One of the simplest root cause analysis techniques is to determine the causes of accidents/incidents at different levels. During any hazard analysis we are always trying to determine the root cause of any accident or incident. Experts who study accidents often do "a breakdown" or analysis of the causes. They analyze them at three different levels:
- direct causes;
- indirect causes;
- basic causes.

(1) **Direct Causes**

When making a detailed analysis of an accident or incident, consider the release of energy and/or hazardous material as a direct cause. Energy or hazardous material is considered to be the force which results in injury or other damage at the time of contact. It is important to identify the direct cause(s). In order to prevent injury, it is often possible to redesign equipment or facilities, and provide personal protection against energy release or contact with hazardous materials.

(2) **Indirect Causes**

Unsafe acts (behavior) and/or unsafe conditions comprise indirect causes of accidents and/or incidents. These indirect causes can inflict injury, property damage, or equipment failure. They allow the energy and/or hazardous material to be released. Unsafe acts can lead to unsafe conditions and vice versa.

When basic causes are eliminated, unsafe acts/unsafe conditions may not occur. (For example: Millie Samuels used a broken ladder because no unbroken ladder existed on the jobsite.) In Millie's case, the basic cause, lack of an unbroken ladder, set up her subsequent unsafe act.

Accidents, thus, have many causes. Basic (root) causes lead to unsafe acts and unsafe conditions (indirect causes). Indirect causes may result in a release of energy and/or hazardous material (direct causes). The direct cause may allow for contact, resulting in personal injury and/or property damage and/or equipment failure (accident).

(3) **Basic Causes**

Some accident investigations result only in the indentification and correction of indirect causes, but indirect causes of accidents are symptoms that some underlying causes exist which are often termed basic causes. By going one step further, accidents can best be prevented by identifying and correcting the basic causes. Basic causes are grouped into policies and decisions, personal factors, and environmental factors.

(4) **Root Cause Analysis**

Root causes are those that, if corrected, would eliminate the accident from occurring again or similar accidents from occurring. They may surround or include several contributing causes. They are a higher order of causes that address a multiple of problems rather than focusing on the single direct cause. An example would be, "Management failed to implement the principles and core functions of a safety and health program. It is management's responsibility to ensure that the workplace has an effective safety and health program and that the workplace is safe for employees to work."

A root cause analysis is not a search for the obvious but an in-depth look at the basic or underlying causes of occupational accidents or incidents. The basic reason for investigating and reporting the causes of occurrences is to enable the identification of corrective actions adequate to prevent recurrence and thereby protect the health and safety of the public, the workers, and the environment. Every root cause investigation and reporting process should include five phases. While there may be some overlap between phases, every effort should be made to keep them separate and distinct. The phases of a root cause analysis are:

Phase I—Data collection

Phase II—Assessment

Phase III—Corrective actions

Phase IV—Inform

Phase V—Follow-up

(5) **Root Cause Analysis Methods**

The most common root cause analysis methods are:

- Events and causal factor analysis identifies the time sequence of a series of tasks and actions and surrounding conditions leading to an occurrence.
- Change analysis is used when the problem is obscure. It is a systematic process that is generally

used for a single occurrence and focuses on elements that have changed.
- Barrier analysis is a systematic process that can be used to identify physical, administrative, and procedural barriers or controls that should have prevented the occurrence.
- Management oversight and risk tree (MORT) analysis is used to identify inadequacies in barriers and controls, specific barrier and support functions and management functions. It identifies specific factors relating to an occurrence and identifies the management factors that permit these risk factors to exist. MORT/Mini-MORT is used to prevent oversight in the identification of causal factors. It lists on the left side of the tree specific factors relating to the occurrence; and on the right side of the tree, it lists the management deficiencies that permit specific risk factors to exist. Management factors support each of the specific barrier and control factors. Included is a set of questions to be asked for each of the barrier and control factors on the tree. As such, they are useful in preventing oversight and ensuring that all potential causal factors are considered. It is especially useful when there is a shortage of experts of whom to ask the right questions. However, because each management oversight factor may apply to specific barrier and control factors, the direct linkage or relationship is not shown but is left up to the analyst. For this reason. Events and causal factor analysis and MORT should be used together for serious occurrences: one to show the relationship, the other to prevent oversight. A number of condensed versions of MORT, called Mini-MORT, have been produced. For a major occurrence justifying a comprehensive investigation, a full MORT analysis could be performed while Mini-MORT would be used for most other occurrences.
- Human performance evaluation identifies factors that influence task performance. The focus of this analysis method is on operability, work environment, and management factors. Man-machine interface studies are frequently done to improve performance. This takes precedence over disciplinary measures. Human performance evaluation is used to identify factors that influence task performance. It is most frequently used for man-machine interface studies. Its focus is on operability and work environment, rather than training of operators to compensate for bad conditions. Human performance evaluations may be used to analyze most occurrences, since many conditions and situations leading to an occurrence have ultimately originated from some task performance problem that results from management planning, scheduling, task assignment, maintenance, and/or inspections. Training in ergonomics and human factors is needed to perform adequate human performance evaluations, especially in man-machine interface situations.
- Kepner-Tregoe problem solving and decision making provides a systematic framework for gathering, organizing, and evaluating information and applies to all phases of the occurrence investigation process. Its focus on each phase helps keep them separate and distinct. The root cause phase is similar to change analysis. Kepner-Tregoe is used when a comprehensive analysis is needed for all phases of the occurrence investigation process. Its strength lies in providing an efficient, systematic framework for gathering, organizing and evaluating information and consists of four basic steps:

1) Situation appraisal to identify concerns, set priorities, and plan the next step.
2) Problem analysis to precisely describe the problem, identify and evaluate the causes and confirm the true cause.
3) Decision analysis to clarify purpose, evaluate alternatives, assess the risks of each option and make a final decision.

4) Potential problem analysis to identify safety degradation that might be introduced by the corrective action, identify the likely causes of those problems, take preventive action and plan contingent action. This final step provides assurance that the safety of no other system is degraded by changes introduced by proposed corrective actions.

These four steps cover all phases of the occurrence investigation process. Thus, Kepner-Tregoe can be used for more than causal factor analysis. Separate worksheets (provided by Kepner-Tregoe) provide a specific focus on each of the four basic steps and consist of step-by-step procedures to aid in the analyses. This systematic approach prevents overlooking any aspect of concern.

An analysis of an accident does not stop with the identification of the direct, indirect, and basic (root) causes of the accident or incident. In order to make positive gains from the event, changes should be made in the interaction of men, machines, materials, methods, and physical and social environments. These changes should result from the recommendations which are derived from the causes identified during the investigation. The goal of these changes is the prevention of future accidents and/or incidents similar to the one investigated.

New Words and Expressions

expeditiously [ˌekspɪˈdɪʃəsli]　　adv. 迅速地，敏捷地
foster [ˈfɒstə]　　v. 鼓励
methodology [ˌmeθəˈdɒlədʒi]　　n. 方法学，方法论
oversight [ˈəʊvəsaɪt]　　n. 失察，疏忽
diffused [dɪˈfjuːzd]　　adj. 散布的；普及的；扩散的
illustrate [ˈɪləstreɪt]　　v. 图解，加插图于；阐明
delineate [dɪˈlɪnɪeɪt]　　v. 描绘
chronologically [ˌkrɒnəˈlɒdʒɪkəli]　　adv. 按年代顺序排列地
validate [ˈvælɪdeɪt]　　v. 确认，证实，验证
inflict [ɪnˈflɪkt]　　v. 造成
underlying [ˌʌndəˈlaɪɪŋ]　　adj. 根本的；潜在的
overlap [ˌəʊvəˈlæp]　　n. 重叠部分
obscure [əbˈskjʊə]　　adj. 模糊的；晦涩的
condensed [kənˈdenst]　　adj. 浓缩的，扼要的
disciplinary [ˈdɪsɪplɪnəri]　　adj. 训练的，规律的
appraisal [əˈpreɪzəl]　　n. 评价，鉴定
contingent [kənˈtɪndʒənt]　　adj. 可能发生的

Unit Five

What Is an OHSMS?

What is an occupational health and safety management system (OHSMS)? One difficulty in evaluating the effectiveness of OHSMS lies in the different meanings given to the term. Finding agreement upon criteria for effectiveness, or methods of measurement and evaluation is especially hard where basic disagreement exists upon what an OHSMS is.

1. The General Characteristics of an OHSMS

All OHSMS owe something to the legacy of general systems theory. Systems theory suggests that there should be four general requirements for an OHSMS, although how these requirements are met in practice allows for considerable diversity. The four general requirements are as follows:
- System objectives.
- Specification of system elements and their inter-relationship; not all systems need have the same elements.
- Determining the relationship of the OHSMS to other systems (including the general management system, and the regulatory system, but also technology and work organisation).
- Requirements for system maintenance (which may be internal, linked to a review phase, or external, linked for example to industry policies that support OHS best practice; system maintenance may vary between systems).

Several Australian authorities upon OHSMS have given definitions broadly consistent with these general system requirements. Thus Bottomley notes what makes an OHSMS a system "is the deliberate linking and sequencing of processes to achieve specific objectives and to create a repeatable and identifiable way of managing OHS. Corrective actions... (are also) central to a systematic approach".

Warwick Pearse also emphasises systemic linkages, defining an OHSMS as "distinct elements which cover the key range of activities required to manage occupational health and safety. These are inter-linked, and the whole thing is driven by feedback loops".

Similarly, Gallagher defines an OHSMS as "... a combination of the planning and review, the management organisational arrangements, the consultative arrangements, and the specific program elements that work together in an integrated way to improve health and safety performance".

2. Voluntary or Mandatory Implementation Methods

One way that OHSMS differ arises from the various methods of implementation. Frick and Wren distinguish three types—voluntary, mandatory and hybrid. Voluntary systems exist where enterprises adopt OHSMS on their own volition. Often this is to implement strategic objectives relating to employee welfare or good corporate citizenship, although there may be other motives such as reducing insurance costs. In contrast, mandatory systems have evolved in a number of European countries where legislation requires adoption of a risk assessment system. Quasimandatory methods may also exist where external commercial pressures take the place of legislative requirements. Thus many businesses adopt OHSMS to comply with the requirements of customers and suppliers, principal contractors and other commercial bodies. Hybrid methods are said to entail a mixture of voluntary motives and legislative requirements.

3. Management Systems or Systematic Management

Following from their distinction between voluntary and mandatory OHSMS, Frick and Wren also separate occupational health and safety "management systems", and the "management systems" of occupational health and safety. Specifically, the former have been characterised as: market-based, promoted typically by consulting firms, and with usually highly formalised prescriptions on how to integrate OHSM within large and complex organisations and also comprehensive demands on documentation.

This "management systems" form must meet stringent criteria. Where these requirements of a "systems" are not met, then the term is said to be inapplicable. On the other hand, "systematic management" is described as "... a limited number of mandated principles for a systematic management of OHS, applicable to all types of employers including the small ones".

This approach stems from methods of regulation found in Europe as well as Australia, where businesses, including smaller ones, are encouraged or required to comply with a less demanding framework than "management systems". One example of this simpler regulatory framework might be the risk assessment principles within the 1989/391 European Union Framework Directive.

Support for such a loose approach to OHSM also exists in Australia. One employer expert on OHS defined systems simply as "just a word for what you do to manage safety". Consistent with this is Bottomley's all-encompassing approach which allows that "... an OHSMS can be simple or complex, it can be highly documented or sparingly described, and it can be home grown or based on an available model".

An example of a relatively simple "systematic" approach to the management of occupational health and safety is to be found in "Small Business Safety Solutions" —a booklet for small business published by the Australian Chamber of Commerce and Industry.

This advocates a four-step process as follows:

Step 1: Commitment to a Safe Workplace (framing a policy based on consultation).

Step 2: Recognising and Removing Dangers (using a danger identification list).

Step 3: Maintaining a Safe Workplace (including safety checks, maintenance, reporting dangers, information and training, supervision, accident investigation, and emergency planning).

Step 4: Safety Records and Information (including records and standards required to be kept by law).

It is debatable whether such a framework for "systematic management" in a small business can include all the elements of planning and accountability that are essential to a "management system" in a large business.

4. System Characteristics: Managerialist and Participative Models

Within "management systems" two different models can be found. The first variant stems from what Nielsen terms "rational organisation theory" (Taylorist and bureaucratic models of organisation). Rational organisation theory is associated with top down managerialist models of OHSMS such as Du Pont. Some authorities now consider most voluntary systems to be managerialist. Thus Frick. et al. observe that "... most voluntary OHSM systems define top management as the (one and only) actor". Conversely, an alternative participative model of "management systems" can be traced to socio-technical systems theory, which emphasises organisational interventions based on analysis of the inter-relationships of technology, environment, the orientation of participants, and organisational structure.

The strengths of this typology are two-fold. First, it is grounded in the literature that discusses alternative approaches to managing OHS and different control strategies, and it reflects the principal debates in that literature. Second, it can be operationalised through empirical tests to see which type of OHSMS performs best.

The typology also faces a difficulty in the fact that the "safe place control strategy" is mandatory in Australia and should be found in all workplaces. There is not, therefore, a clear choice between two mutually exclusive control strategies; the workplace with dominant safe person characteristics should also be implementing safe place characteristics.

5. Degree of Implementation: Quality Levels

Frick and Wren expand upon their distinction between mandatory and voluntary OHSMS to further identify three levels of systems objectives, drawn from the literature on product quality control, that represent different levels of achievement and measures of OHSM performance.

6. Degree of Implementation: Introductory and Advanced Systems

The idea that there may be different levels of OHSM has been interpreted another way in Australia where performance levels in some programs are explicitly developmental (the business graduating up an ascending ladder as it demonstrates compliance with the requirements of each successive level).

One example of an Australian program with developmental steps is the South Australian Safety Achiever Business System (SABS) (formerly known as the Safety Achiever Bonus Scheme). The Program specifies five standards (commitment and policy, planning, implementation, measurement and management systems review and implementation) linked in a continuous improvement cycle. Three "levels" of implementation are then prescribed cumulatively introducing all five standards from a basic or introductory program to a proven continuous improvement system. Different evaluation standards are prescribed for each level.

7. OHSMS Diversity and Evaluation: A Summary

While, in general, this Report advocates care in defining OHSMS with respect to the problems outlined above, for the purpose of this project an inclusive approach to the phenomena is to be adopted. In particular, the term OHSMS will be used broadly to encompass both the highly complex formal systems adopted voluntarily by some businesses as well as the more rudimentary mandatory or advisory frameworks offered to and implemented by small business.

So far, we have shown that OHSMS can vary upon a number of dimensions relating to method of implementation, system characteristics, and degree of implementation. Such variance is important because it affects evaluation and measurement of OHSMS performance. Measures appropriate for one dimension of a system will be irrelevant to another. Evaluation of OHSMS effectiveness may need to take account of what systems are expected to do. Are they to meet complex system or simple design standards? Are they implemented at the behest of management or external OHS authorities? Are objectives the simple ones such as reducing direct lost-time injuries or do they include satisfying multiple stakeholders? Are they at an early or established stage of development; and which of several different configurations of control strategy and management structure/style is adopted?

Drawing upon the review above, the diagram below sets out five key dimensions on which OHSMS vary that need to be considered in evaluation and measurement.

8. OHSMS Diversity: 5 Key Dimensions for Evaluation

While all systems must meet the general requirements for an OHSMS, diversity may occur along five key dimensions as follows:
- implementation method (voluntary, mandatory or hybrid);
- control strategy (safe person/safe place);
- management structure and style (innovative or traditional);
- degree of implementation (from meeting basic specifications to meeting stakeholder needs);
- degree of implementation (from introductory stage to fully operational).

New Words and Expressions

diversity [daɪˈvɜːsɪti]　　　　　　　　n. 差异，多样性
deliberate [dɪˈlɪbəreɪt]　　　　　　　adj. 深思熟虑的，精心策划的
planning and review　　　　　　　　计划与评审
hybrid [ˈhaɪbrɪd]　　　　　　　　　　n. 混合物，混杂物
volition [vəʊˈlɪʃən]　　　　　　　　　n. 意志
entail [ɪnˈteɪl]　　　　　　　　　　　v. 遗留给，传给（弊害等）
prescription [prɪˈskrɪpʃən]　　　　　n. 规定，法规
stringent [ˈstrɪndʒənt]　　　　　　　adj. 严格的
mandated [ˈmændeɪtɪd]　　　　　　　adj. 委托管理的，托管的

advocate [ˈædvəkɪt]	v. 主张，提倡
planning and accountability	计划与职责
intervention [ˌɪntəˈvenʃən]	n. 干预，介入
typology [taɪˈpɒlədʒi]	n. 类型，种类
encompass [ɪnˈkʌmpəs]	v. 包含，涵盖
rudimentary [ˌruːdɪˈmentəri]	adj. 未发展的；初步的，根本的
irrelevant [ɪˈrelɪvənt]	adj. 不恰当的，不对题的
behest [bɪˈhest]	n. 要求，指令
stakeholder [ˈsteɪkhəʊdə]	n. 风险金管理机构

Translation Skill

科技英语翻译技巧（四）——句子成分转换

句子成分转换是指词类不变而成分改变的译法。通过改变原文中某些句子成分，以达到译文逻辑正确、通顺流畅、重点突出等目的。

一、谓语、表语、宾语、状语转译成主语

Gases *differ* from liquids in that the former have greater compressibility than the latter.
气体和固体的*区别*，在于前者比后者有更大的压缩性。（谓语转译成主语）

Rubber is a better *dielectric* but a poorer *insulator* than air.
橡胶的*介电性*比空气好，但*绝缘性*比空气差。（表语转译成主语）

Water has a *density* of 62.4 pounds per cubic foot.
水的*密度*是每立方英尺 62.4 磅。（宾语转译成主语）

Aluminum is very light *in weight*, being only one third as heavy as iron.
铝的*重量*很轻，只有铁的 1/3。（状语转译成主语）

The same signs and symbols of mathematics are used *throughout the world*.
*全世界*都使用同样的数学记号和符号。（状语转译成主语）

二、主语、定语、表语、状语转译成谓语

The *prevention* of many types of occupational diseases is distinctly within the realm of possibility.
许多职业病显然可以*预防*。（主语转译成谓语）

Also *present* in solids are numbers of free electrons.
固体中也*存在*着大量的自由电子。（表语转译成谓语）

A semiconductor has a *poor* conductivity at room temperature, but it may become a good conductor at high temperature.（定语转译成谓语）
在室温下，半导体电导率*差*，但在高温下，它可能成为良导体。

The wide applications of computers affect *tremendously* the development of science and technology. （状语转译成谓语）
计算机的广泛应用，对科学技术的影响*极大*。

三、主语、状语、宾语转译成定语

CMOS chips in the computer work a thousand times more rapidly than nerve cells in the human brain.

计算机芯片的工作比人类大脑中的神经细胞要快1000倍。（主语转译成定语）

In Britain the first stand-by gas-turbine electricity generator was in operation in Manchester in 1952.

英国的第一台辅助燃气发电机于1952年在曼彻斯特开始运转。（状语转译成定语）

By 1914 Einstein had gained *world fame*.

1914年爱因斯坦已成了世界著名的科学家。（宾语转译成定语）

四、定语、谓语、主语转译成状语

It was an amazing piece of *scientific* clairvoyance, comparable perhaps to Charles Babbage's anticipation of the principle of the computer.

这个理论在科学上充满了远见卓识，也许可以跟巴贝奇的计算机原理相提并论。（定语转译成状语）

After more experiments, Galileo *succeeded* in making a much better telescope.

又做了一些实验之后，伽利略成功地制造了一架好得多的望远镜。（谓语转译成状语）

The result of his revolutionary design is that the engine is much smaller, works more smoothly, and has fewer moving parts.

由于他在设计上的革新，发动机变得小多了，工作得更平稳了，活动部件也少了。（主语转译成状语）

五、主语、状语、谓语转译成宾语

The mechanical energy can be changed back into electrical energy by means of a generator.

利用发电机可以把机械能重新转变成电能。（主语转译成宾语）

Television has been *successfully* sent by laser, too.

用激光发射电视也获得了成功。（状语转译成宾语）

The sun *affects* tremendously both the mind and body of a man.

太阳对人的身体和精神都有极大的影响。（谓语转译成宾语）

六、主语、宾语转译成表语

A great contribution of Edison was the carbon microphone.

炭精传声器是爱迪生的一大贡献。（主语转译成表语）

The production had considerable *difficulty* getting patent protection if it had no patent right.

如果产品没有专利权，要获得专利保护是相当困难的。（宾语转译成表语）

由此看来，句子成分的转译显得变化万千，异彩纷呈，难于穷尽，几乎所有的句子成分都可互相转译。进行成分转译的目的是使译文通顺，合乎汉语习惯，以及更好地与上下文响应。

Reading Material

The Standard for Occupational Health and Safety

An organization's activities can pose a risk of injury or ill-health, and can result in a serious impairment of health, or even fatality, to those working on its behalf. According to an estimate by the International Labor Organization (ILO), there were 2.34 million deaths in 2013 as a result of work activities. The greatest majority (2 million) are associated with health issues, as opposed to injuries. The Institute of Occupational Safety and Health (IOSH) estimates there are 660,000 deaths a year as a result of cancers arising from work activities. Consequently, an organization is responsible for ensuring that it minimizes the risk of harm to the people that may be affected by its activities (e.g. its workers, its managers, contractors, or visitors), and particularly if they are engaged by the organization to perform those activities as part of their occupation. It is important for the organization to eliminate or minimize its occupational health and safety (OHS) risks by taking appropriate preventive measures.

1. OHSAS 18000 Series

Before 1999, organizations worldwide recognize the need to control and improve health and safety performance and do so with occupational health and safety management system (OHSMS). There was an increase of national standards and proprietary certification schemes to choose from. Organizations had to choose from a range of national health and safety standards and proprietary certification schemes. But this led to confusion and fragmentation in the market, while undermining the credibility of individual schemes and creating potential trade barriers. Recognizing this deficit, an international collaboration called the Occupational Health and Safety Assessment Series (OHSAS) Project Group was formed to create a single unified approach. The Group comprised representatives from national standards bodies, academic bodies, accreditation bodies, certification bodies and IOSH, with the UK's national standards body, BSI Group, providing the secretariat. Drawing on the best of existing standards and schemes, the OHSAS Project Group published the OHSAS 18000 Series in 1999. It lays out all the elements that can be integrated with other management systems to boost a company's occupational health and safety performance. The Series consisted of two specifications: 18001 provided requirements for an OHS management system and 18002 gave implementation guidelines. In 2005, around 16,000 organizations in more than 80 countries were using the OHSAS 18001 specification. By 2009 more than 54,000 certificates had been issued in 116 countries to OHSAS or equivalent OHSMS standards.

In July 2007, the OHSAS 18001 specification was updated and more closely aligned with the structures of other management system standards such as the ISO 14001. This helped organizations to bring their existing management systems more easily in line with the standard. Additionally, the "health" component of "health and safety" was given greater emphasis. The UK then decided to adopt OHSAS 18001 as a British Standard and created BS OHSAS 18001. So BS OHSAS 18001 is designed to help organizations implement a framework that identifies and controls health and safety risks, reduces potential accidents, aids legislative compliance and improves overall performance. The standard also demonstrates how to develop and implement a policy with the right objectives for organizations of all types and sizes, covering diverse geographical, cultural and social conditions. BS OHSAS 18001 is a widely recognized and popular occupational health and safety management system, and it is an internationally applied occupational health and safety management system. BSI Group subsequently adopted the updated 18002 guidance specification for publication as BS OHSAS 18002 in 2008.

2. ISO 45001: Occupational Health and Safety Management Systems—Requirements

The issue of work-related injuries and diseases is significant and growing, both for employers and the economy. To combat this problem, in October 2013 a project committee, ISO PC 283, met in London to create the first working draft of ISO 45001 (occupational health and safety management systems—requirements) which was expected to help organizations reduce this burden globally by providing a framework to improve employee safety, reduce workplace risks and create better, safer working conditions. The standard is currently being developed by a committee of occupational health and safety experts and will follow other standard management system protocols approaches such as ISO 14001 and ISO 9001. It will take into account other International Standards in this area such as OHSAS 18001, the International Labour Organization's ILO-OSH Guidelines, various national standards and the ILO's international labor standards and conventions. The standard is targeted to be published by February of 2018.

ISO 45001 is an International Standard that specifies requirements for an OHS management system, with guidance for its use, to enable an organization to proactively improve its OHS performance in preventing injury and ill-health. ISO 45001 is intended to be applicable to any organization regardless of its size, type and nature. All of its requirements are intended to be integrated into an organization's own management process.

There is a greater focus on top management to demonstrate leadership and commitment with respect to the management system and to ensure consultation and participation of workers in the development, planning, implementation and continual improvement of the OHS management system. Top management have a responsibility to ensure that the importance of effective OHS management is communicated and understood by all parties and ensuring that the OHS management system achieves its intended outcomes.

The introduction of a more holistic risk and opportunity management into the OHS management system now reinforces its use as a governance tool. It will enable the identification of opportunities that contribute to further improvement in OHS performance and improved worker safety. Organizations will improve their ability to identify and manage risks more effectively making it more resilient. ISO 45001 is based on Annex SL—the new ISO high level structure (HLS) that brings a common framework to all management systems. This helps to keep consistency, align different management system standards, offer matching sub-clauses against the top-level structure and apply common language across all standards. With the new standard in place, organizations will find it easier to incorporate their OHS management system into the core business processes and get more involvement from senior management.

It is logical that those working closest to an OHS risk will be knowledgeable about it. As such, the participation of workers in the establishment, implementation and maintenance of an OHS management system can play an important role in ensuring that the risks are managed effectively. ISO 45001 emphasizes the need for worker participation in the functioning of an OHS management system, as well as requiring that an organization ensures that its workers are competent to do their assigned tasks safely

3. What will be the benefits of using ISO 45001?

An ISO 45001 based the OHS management system will enable an organization to improve its OHS

performance by:
- developing and implementing an OHS policy and OHS objectives;
- establishing systematic processes which consider its "context" and which take into account its risks and opportunities, and its legal and other requirements;
- determining the hazards and OHS risks associated with its activities; seeking to eliminate them, or putting in controls to minimize their potential effects;
- establishing operational controls to manage its OHS risks and its legal and other requirements;
- increasing awareness of its OHS risks;
- evaluating its OHS performance and seeking to improve it, through taking appropriate actions;
- ensuring workers take an active role in OHS matters.

In combination these measures will ensure that an organization's reputation as a safe place to work will be promoted, and can have more direct benefits, such as:
- improving its ability to respond to regulatory compliance issues;
- reducing the overall costs of incidents;
- reducing downtime and the costs of disruption to operations;
- reducing the cost of insurance premiums;
- reducing absenteeism and employee turnover rates;
- recognition for having achieved an international benchmark (which may in turn influence customers who are concerned about their social responsibilities).

While the standard requires that OHS risks are addressed and controlled, it also takes a risk based approach to the OHS management system itself, to ensure: ① that it is effective and ② being improved to meet an organization's ever changing "context". This risk based approach is consistent with the way organizations manage their other "business" risks and hence encourages the integration of the standards requirements into organizations' overall management processes.

New Words and Expressions

International Labor Organization (ILO)	国际劳工组织
occupational health and safety	职业健康与安全
health and safety performance	健康与安全绩效
occupational health and safety management system (OHSMS)	职业健康与安全管理体系
fragmentation [ˌfræɡmenˈteɪʃ(ə)n]	n. 分裂,破碎
holistic [həʊˈlɪstɪk]	adj. 全盘的,整体的;功能整体性
impairment [ɪmˈpeəm(ə)nt]	n. 损害,损伤

Unit Six

Industrial Hygiene

Industrial hygiene has been defined as "that science or art devoted to the anticipation, recognition, evaluation, and control of those environmental factors or stresses, arising in or from the workplace, which may cause sickness, impaired health and well-being, or significant discomfort and inefficiency among workers or among the citizens of the community".

The industrial hygienist, although basically trained in engineering, physics, chemistry, or biology, has acquired by undergraduate and/or postgraduate study and experience, a knowledge of the effects upon health of chemical and physical agents under various levels of exposure. The industrial hygienist is involved with the monitoring and analytical methods required to detect the extent of exposure, and the engineering and other methods used for hazard control. The industrial hygienist looks at specific environmental factors (stresses) or hazards. These factors are physical, biological, ergonomic and chemical.

1. Physical Hazards

Physical hazards include excessive levels of nonionizing and ionizing radiation, noise, vibration, and extremes of temperature and pressure. Any of these have or can have serious adverse effects upon your workforce. You should identify any of these which exist in your work environment and which present a risk to your employees.

Physical hazards are defined as those type of hazards that can cause harm to a worker from an external source. Types of physical hazards are loud noise (equipment), temperature extremes (working in personal protective equipment), radiation (exposures to the infrared or gamma rays), chemical bum (acids or caustics), fire and/or explosions. Other physical hazards include, but are not limited to, slips and falls, exposed machinery because of improper guarding, live electrical circuits or conductors, equipment moving about on site, confined spaces, and falling objects.

Noise is a serious hazard when it results in temporary or permanent hearing loss, physical or mental disturbance, any interference with voice communications, or the disruption of a job, rest, relaxation, or sleep. Noise is any undesired sound and is usually a sound that bears no information with varying intensity.

It interferes with the perception of wanted sound, and is likely to be harmful, cause annoyance, and/or interfere with speech.

The noise created by circular saws, planers, or high speed grinders and similar power tools is narrow band noise. This high frequency type of noise is very damaging to the inner ear. Impulse type noise is generated by energy bursts occurring repetitively or one at a time. Noise from a jack hammer is an example of repetitive impulse noise. The firing of a gun is an example of a singular impulse noise. All types of noise can harm you if it is high intensity, and/or the exposure time is prolonged or repetitive.

A healthy young person can detect sounds in the 20 to 20,000 cycles per second range. As aging takes place, some hearing is lost. Higher frequencies cause the most damage to our ears and most people who have hearing loss have high frequency losses first. Loudness or softness is determined by the intensity or sound pressure. The more power driving the sound, the higher the pressure. This is measured with an instrument called a sound level meter (SLM) in units called decibels (dB). Sounds that can just be heard by a person with very good hearing in an extremely quiet location are assigned the value of 0 dB. Ordinary speech is around 50 to 60 dB. At about 120 dB the threshold of pain is reached. This would be like hearing a jet engine about 50 feet away.

Noise dose limits are now required for workplaces to minimize hearing loss from occupational exposure. Although louder noise is allowed for brief periods during the workday, the mandatory noise level limit (set by OSHA), is 90 dBA. An employer must make hearing protection available, provide training, and provide hearing tests when the noise level exceeds 85 dBA, time-weighted average. As a basic rule, if you cannot hear the snap of your fingers at arms length you should be using hearing protection. Over 90 dBA, employers must assure that protection is being used. Heat stress is a serious physical hazard that should always be considered on a construction job site especially during the summer months. The chance of developing heat stress increases with increased humidity, hot environments, and the use of personal protective equipment. Sweating is the most effective means of losing excess heat, as long as adequate fluids are taken in to replace the sweat. When individuals are severely stressed by the heat, they may stop sweating with the most severe consequences of heat stress occurring. Adequate rest periods, availability of large amounts of replacement fluids, and frequent monitoring are essential to prevent the consequences of heat stress which may occur without warning symptoms. The body maintains a normal temperature (98.6° F) in a hot environment by two methods:

- Sending more blood to the skin.
- Sweating.

Cold stress occurs when temperatures go down, the body maintains its temperature by reducing blood flow to the skin. This causes a marked decrease in skin temperature. The most extreme effect is on the extremities (fingers, toes, earlobes and nose). When hands and fingers become cold they become numb and insensitive, and there is an increased possibility of accidents. If the restriction of blood flow to the skin is not adequate to maintain temperature then shivering occurs. If this is not adequate to warm the body, then a marked decrease in temperature (hypothermia) may occur. Workers that may be at increased risk are:

- Doing hard labor who become fatigued and/or wet either from sweating or contact with water. Taking sedatives or drinking alcohol before or during work. Workers with chronic diseases that affect the heart and/or blood vessels of the hands or feet. Not physically fit or have not worked in a cold environment recently.

- Those who use pavement breakers or other vibrating equipment.

Radiation is divided into two major categories, based its effect on living tissue: ①non-ionizing; ②ionizing radiation. Ionizing radiation has the ability to change or destroy the atomic (chemical) structure of cells, non-ionizing radiation does not. Some types of nonionizing radiation that we are exposed to everyday include microwave energy used for cooking, and radio waves used in broadcasting over radio and television. Types of ionizing radiation we are exposed to are cosmic rays from the sun and stars, terrestrial radiation from the earth, nuclear radiation from reactors, and medical radiation from x-rays.

Although non-ionizing radiation is not as hazardous as ionizing, there are exposures that can cause severe injuries. Non-ionizing radiation is generated by such things as the sun, lamps, welding arcs, lasers, plastic sealers, and radio or radar broadcast equipment. Since the eye is 'the primary organ at risk to all types of non-ionizing radiation, eye protection is very important. Protective glasses should be selected based on the type of radiation exposure, for example, sunlight or welding flashes. Ionizing radiation is so named because it has enough energy to change (ionize) atoms and molecules, the building blocks of all matter. There are four natural types of ionizing radiation—alpha and beta particles, gamma rays, and neutrons.

Vibration is a much more difficult physical factor to address since it is often difficult to attach the symptoms with the exposure. Also, our ability to measure vibration and determine what measurements will cause ill effects to workers is very limited.

2. Biological Hazards

Biological hazards include vermin, insects, molds, fungi, viruses, and bacterial contaminants. Items such as sanitation and housekeeping items regarding potable water, removal of industrial waste and sewage, food handling, and personal cleanliness have the potential to exacerbate the potential risk of biological hazards. Biological agents may be a part of the total environment or may be associated with certain occupations such as agriculture. Biological agents in the workplace include viruses, rickettsia (organisms that cause diseases), bacteria, and parasites of various types. Diseases transmitted from animal to man are common. Infections and parasitic diseases may also result from exposure to insects or by drinking contaminated water. Exposure to biohazards may seem obvious in occupations such as nursing, medical research, laboratory work, farming, and handling of animal products (slaughterhouses and meat packing operations). The sting of bees, which many workers are allergic to, is not so obvious a biological hazard. Biohazards may be transmitted to a person through inhalation, injection, ingestion or physical contact. Many plants and animals produce irritating, toxic, or allergenic (causing allergic reactions) substances. Dusts may contain many kinds of allergenic materials, including insect scale, hairs, and fecal dust, sawdust, plant pollens, and fungal spores. Other hazards include bites or attacks by domestic and wild animals. Workers on hazardous waste sites may risk exposure to bites from venomous snakes or poisonous spiders.

3. Ergonomic Hazards

Ergonomic hazards include improperly designed tools or work areas. Improper lifting or reaching, poor visual conditions, or repeated motions in an awkward position can result in accidents or illnesses in the occupational environment. Designing the tools and the job to be done to fit the worker should be of prime importance. When repetitive motion injuries occur, they often result from continuous use of a body part often in an unnatural posture employing more force than is normal for the body part. This may result in irritation, fluid build up, or thickening of the tendons and ligaments in the wrists, or damage to nerves or blood vessels. Severe pain may occur along with numbness, and loss of movement may occur. Weakness of the hand, arm or other body part may occur, making it difficult to hold objects and perform grasping motions. The worker may drop objects, be unable to use keys, or count change because of these injuries. Surgical treatment may be necessary if the symptoms are severe and if other measures do not provide relief. Other ergonomic hazards include manual handling of objects and materials where lifting and carrying are done. Lifting is so much a part of many everyday jobs that most of us do not think about it. But it is often done wrong, with unfortunate results such as pulled muscles, disk injuries, or painful hernias. Intelligent application of engineering and biomechanical principles is required to eliminate hazards of this kind.

4. Chemical Hazards

Chemical hazards arise from excessive airborne concentrations of mists, vapors, gases, or solids that are in the form of dusts or fumes. In addition to the hazard of inhalation, many of these materials may act as skin irritants or may be toxic by absorption through the skin. There are thousands and thousands of potentially harmful chemicals found in the workplace. Workers face the possibility of exposure on a daily basis to these harmful chemicals. The majority of the occupational health hazards arise from inhaling chemical agents in the form of vapors, gases, dusts, fumes and mists, or by skin contact with these materials. The degree of risk of handling a given substance depends on the magnitude and duration of exposure. To recognize occupational factors or stresses, a health and safety professional must first know about the chemicals used as raw materials and the nature of the products and byproducts manufactured. This sometimes requires great effort. The required information can be obtained from the Material Safety Data Sheet (MSDS) that must be supplied by the chemical manufacturer or importer to the purchaser for all hazardous materials under the Hazard Communication Standard (HCS). The MSDS is a summary of the important health, safety, and toxicological information on the chemical or the mixture ingredients. Other stipulations of the HCS require that all containers of hazardous substances in the workplace be labeled with appropriate warning and identification labels. If the MSDS or the label does not give complete information but only trade names, it may be necessary to contact the manufacturer of the chemicals to obtain this information. Many industrial materials such as resins and polymers are relatively inert and non-toxic under normal conditions of use, but when heated or machined, they may decompose to form highly toxic by-products. Information concerning these types of hazardous products and by-products must also be included in the company's Hazard Communication Program. Breathing of some materials can irritate the upper respiratory tract or the terminal passages of the lungs and the air sacs, depending upon the solubility of the material. Contact of irritants with the skin surface can produce various kinds of dermatitis.

The presence of excessive amounts of biologically inert gases can dilute the atmospheric oxygen below the level required to maintain the normal blood saturation value for oxygen and disturb cellular processes. Other gases and vapors can prevent the blood from carrying oxygen to the tissues or interfere with its transfer from the blood to the tissue, thus producing chemical asphyxia or suffocation. Carbon monoxide and hydrogen cyanide are examples of chemical asphyxiants. Some substances may affect the central nervous system and brain to produce narcosis and/or anaesthesia. In varying degrees, many solvents have these effects.

Substances are often classified according to the major reaction that they produce, as asphyxiants, systemic toxins, pneumoconiosis-producing agents, carcinogens, irritant gases, or high dust levels.

New Words and Expressions

hygiene [ˈhaɪdʒiːn]	n. 卫生学, 保健法; 卫生
ionize [ˈaɪənaɪz]	v. 使离子化; 电离
infrared [ˌɪnfrəˈred]	adj. 红外线的 n. 红外线
caustic [ˈkɔːstɪk]	adj. 腐蚀性的 n. 腐蚀剂
circular saw	圆锯
planer [ˈpleɪnə]	n. 刨工; 刨机
jack hammer	凿岩锤
earlobe [ˈɪələʊb]	n. 耳垂
hypothermia [ˌhaɪpəʊˈθɜːmɪə]	n. 降低体温
sedative [ˈsedətɪv]	n. 镇静剂; 止痛药
pavement breaker	混凝土路面破碎机
cosmic ray	宇宙线, 宇宙射线
terrestrial radiation	地面辐射
welding arc	焊弧
welding flash	弧光灼伤
vermin [ˈvɜːmɪn]	n. 害虫, 寄生虫
fungi [ˈfʌŋgiː]	adj. 真菌类的; 由真菌引起的
sanitation [ˌsænɪˈteɪʃən]	n. 卫生, 卫生设施
exacerbate [eksˈæsə(ː)beɪt]	v. 恶化, 增剧, 使加剧
rickettsia [rɪˈketsɪə]	n. 发疹伤寒等的病原体
parasite [ˈpærəsaɪt]	n. 寄生虫; 食客
slaughterhouse [ˈslɔːtəhaʊs]	n. 屠宰场
allergic [əˈlɜːdʒɪk]	adj. [医] 过敏的, 患过敏症的
inhalation [ˌɪnhəˈleɪʃən]	n. 吸入
ingestion [ɪnˈdʒestʃən]	n. 摄取
irritating [ˈɪrɪˌteɪtɪŋ]	adj. 刺激的
fecal [ˈfiːkəl]	adj. 排泄物的, 渣滓的
sawdust [ˈsɔːdʌst]	n. 锯屑
pollen [ˈpɒlɪn]	n. 花粉
spore [spɔː]	n. 孢子
venomous [ˈvenəməs]	adj. 有毒的, 分泌毒液的
irritation [ˌɪrɪˈteɪʃən]	n. 发炎, 疼痛
ligament [ˈlɪgəmənt]	n. 系带, 韧带
hernia [ˈhɜːnjə]	n. [医] 疝气, 脱肠
toxicological [ˌtɒksɪˈkɒlədʒɪkəl]	adj. 毒物学的
resin [ˈrezɪn]	n. 树脂
respiratory tract	呼吸道

air sac	气囊
solubility [ˌsɒljʊˈbɪlɪti]	n. 溶度，溶性，溶解性
dermatitis [ˌdɜːməˈtaɪtɪs]	n. [医] 皮炎
asphyxia [æsˈfɪksɪə]	n. 窒息，昏厥
suffocation [ˌsʌfəˈkeɪʃn]	n. 窒息
carbon monoxide	n. [化] 一氧化碳
cyanide [ˈsaɪənaɪd]	n. [化] 氰化物
narcosis [nɑːˈkəʊsɪs]	n. 昏迷状态，麻醉
anaesthesia [ˌænəsˈθiːzjə]	n. [医] 感觉缺乏，麻木，麻醉（法）
solvent [ˈsɒlvənt]	n. 溶媒，溶剂
toxin [ˈtɒksɪn]	n. [生化] 毒素
pneumoconiosis [ˈnjuːməˌkəʊnɪˈəʊsɪs, ˈnuː-]	n. [医] 肺尘症
carcinogen [kɑːˈsɪnədʒən]	n. 致癌物质

Translation Skill

科技英语翻译技巧（五）——常见多功能词的译法（I）

多功能词 as，it，that，what 在科技文章中广泛使用，出现频率很高，有必要提出来重点讨论。

一、as 的译法

1. as 作为介词的译法

as 作为介词可引出主语补足语、宾语补足语、状语、同位语等，可译为"作为，为，以，是，当作"等。

Fire protection engineers define the term explosion *as* an "effect" produced by a sudden violent expansion of gases.

消防工程师把爆炸定义为气体瞬时剧烈膨胀的效应。

We take steel *as* the leading engineering material.

我们把钢作为主要的工程材料。

2. as 作为关系代词的译法

as 作为关系代词可以单独使用，也可以与 such，the same 等词搭配使用，引导定语从句。

1) as 单独使用，引导定语从句或省略定语从句时，可译为"正如，如"或"这，这样"等，有时 as 可略去不译。

As known to us, inertia is an absolute quality possessed by all bodies.

正如我们所知，惯性是所有物体都具有的一种绝对属性。

As have been found there are more than a hundred elements.

已经发现，元素有100多种。（略去不译）

2) such...as，such as 引出的定语从句，可译为"像……这（那）样的"，"那样（种）的，像……之类的"等。

Fire deals with *such* ideas as substance, temperature, explosion and combustion.

火灾涉及诸如物质、温度、爆炸和燃烧之类的概念。

3）the same...as，the same as 引出的定语从句，可译为"和……一样"，与……相同，和……相等"。

The weight of an object on the moon is *not the same* as its weight on the earth.

某一物体在月球上的重量与该物体在地球上的重量不一样。

3. as 作为关系副词的译法

as 作为关系副词引导定语从句的译法，与作为关系代词引导定语从句的译法基本相同。

A vector multiplied by a scalar quantity is a vector in the same direction *as* the original vector.

乘以标量的矢量是一个与原矢量同向的矢量。

4. as 作为连词的译法

as 作为连词，可引导时间、原因、比较、让步等状语从句。不同的状语从句中的 as 有不同的译法，分述于后。

1）as 引导时间状语从句，可译为"（当）……时，随着……"等。

The force of gravitational attraction between two bodies decrease *as* the distance between them increases.

两物体之间的引力随着两物体之间的距离增大而减小。

2）as 引导原因状语从句，可译为"由于……，因为……所以……"等。

As liquids and gases flow, they are called fluids.

由于液体和气体能流动，因此称为流体。

3）as 引导比较状语从句时，as + 形容词或副词 + as 可译为"与……一样（同样）……"; not so + 形容词或副词 + as；"不如（没有）……（那样）……"; as + 形容词或副词 + as 可译为"又……又……"; as long as 可译为"多久……多久"。

The wheel turns *as* fast *as* stable.

这只轮子旋转得又快又稳。

The development stage lasts *as* long *as* it needs to.

研制阶段需要持续多久，就持续多久。

4）as 引导让步状语从句，可译为"虽然……，尽管……"。

Complicated *as* the problem is, it can be worked out in a few minutes by a computer.

尽管那个问题很复杂，计算机能在几分钟内将它解出。

5）as 引导方式状语从句，可译为"像……那样，正如……，正如……一样"等。

Friction, *as* the term is understood in mechanics, is the resistance to relative motion between two bodies in contact.

正如大家在力学中所知的，摩擦就是两个相接触的物体产生相对运动的阻力。

二、it 的译法

1. it 的省略

it 用作无人称主语，表示天气、时间、距离等；或用作先行主语、先行宾语，可略去不译。

We found *it* necessary to use modern composite materials for tall buildings.

我们认为，采用现代复合材料建造高层建筑是必要的。

2. it 引导强调句型的译法

在这种句型中，it 本身无词汇意义，但在被强调部分的前面加译"是""正是""只有"

等字。

It is to reduce accidents that safety valves are designed.

正是为了减少事故，才设计了安全阀。

3. it 用作人称代词的译法

1）it 代替上文所提及之事物，一般可译成"它"。

A new science, safety behavior, has emerged, and *it* is very closely allied to management and psychology.

出现了一门新兴学科——安全行为学，它与管理学和心理学关系密切。

2）英语中的多数状语从句可位于主句之前或之后，因此指代的某一事物在主句中可用名词也可用代词；但汉语中的状语从句一般位于主句之前，而指代某一事物总是先出现名词后使用代词。基于上述情况，it 往往译成所代替的名词。

The light is slowed down as *it* goes through the lens.

当光线穿入透镜时，速度就减慢下来。

3）it 代替上文所提及的某件事或上面的整个句子时，可译成"这，这一点"。

Is a structure safe when loaded? *It* is a problem to be studied.

结构受载后还安全吗？这是一个需要研究的问题。

Reading Material

Occupational Illness

Occupational illnesses are not as easily identified as injuries. According to the Bureau of Labor Statistics there were 5.7 million injuries and illnesses reported in 1999. Of this number only 372,000 cases of occupational illnesses were reported. The 372,000 occupational illnesses included repeat trauma such as carpal tunnel syndrome noise-induced hearing loss, and poisonings. It is my professional opinion that many occupational injuries go unreported when the employer or worker is not able to link exposure with the symptoms the employees are exhibiting. Also, physicians fail to ask the right questions regarding the patient's employment history, which can lead to the commonest diagnosis of a cold or flu. This has become very apparent with the recent occupational exposure to anthrax where a physician sent a worker home with anthrax without addressing potential occupational exposure hazards. Unless the physician is trained in occupational medicine, he or she seldom addresses work as the potential exposure source.

This is not entirely a physician problem by any means since the symptoms which are seen by the physician are often those of flu and other common illnesses suffered by the general public. It is often up to the employee to make the physician aware of on-the-job exposure. If you notice, I have continuously used the term exposure since, unlike trauma injuries and deaths, which are usually caused by the release of some source of energy, occupational illnesses are often due to both short-term and long-term exposures. If the results of an exposure lead to immediate symptoms, it is said to be acute. If the symptoms come at a later time, it is termed a chronic exposure. The time between exposure and the onset of symptoms is called the latency period. It could be days, weeks, months, or even years, as in the case of asbestos where asbestosis or lung cancer appears 20 to 30 years after exposure. Looking at a large number of death certificates (20,000) from specific groups of workers, you often see a significant number of cases of specific types of cancer such as liver, thyroid, or pancreatic cancer that do not appear in the same number in the normal adult population. This leads one to believe that something the workers were exposed to in the work environment may have caused their demise.

It is often very difficult to get employers, supervisors, and employees to take seriously the exposures in the workplace as a potential risk to the workforce both short and long term, especially long term. "It can't be too bad if I feel all right now." This false sense of security is illustrated by the 90,000 occupational illness deaths which are estimated by the Bureau of Labor Statistics to occur each year. This far surpasses the 6,000 occupation trauma deaths a year. If both trauma and illness deaths are added together it would be equivalent to the lives lost to a jumbo jet crashing every day of the year. Would an aviation record like this be acceptable to you? If not, I doubt that you would be flying. It is time for employers and the workforce to take on-the-job exposures as a potentially serious threat.

1. Identifying Health Hazards

Health-related hazards must be identified (recognized), evaluated, and controlled in order to prevent occupational illnesses which come from exposure to them. Health-related hazards come in a variety of forms, such as chemical, physical, ergonomic, or biological:

- Chemical hazards arise from excessive airborne concentrations of mists, vapors, gases, or solids that are in the form of dusts or fumes. In addition to the hazard of inhalation, many of these materials may act as skin irritants or may be toxic by absorption through the skin. Chemicals can also be ingested, although this is not usually the principal route of entry into the body.
- Physical hazards include excessive levels of nonionizing and ionizing radiations, noise, vibration, and extremes of temperature and pressure.
- Ergonomic hazards include improperly designed tools or work areas. Improper lifting or reaching, poor visual conditions, or repeated motions in an awkward position can result in accidents or illnesses in the occupational environment. Designing the tools and the job to be done to fit the worker should be of prime importance. Intelligent application of engineering and biomechanical principles is required to eliminate hazards of this kind.
- Biological hazards include insects, molds, fungi, viruses, vermin (birds, rats, mice, etc.) and bacteria contaminants (sanitation and housekeeping items such as potable water, removal of industrial waste and sewage, food handling, and personal cleanliness can contribute to the effects from biological hazards). Biological and chemical hazards can overlap.

These health-related hazards can often be difficult and elusive to identify. A common example of this is a contaminant in a building which has caused symptoms of illness. Even the evaluation process may not be able to detect the contaminant which has dissipated before a sample can be collected. This leaves nothing to control and possibly no answer to what caused the illnesses.

Most skin disorders can be prevented with the proper use of personal protective equipment (PPE) and good personal hygiene (washing hands, etc.). Usually skin disorders are caused by exposure to chemical and result in nothing more than a rash that is cured by proper PPE or removal or substitution (using a safer chemical) of the chemical. Some skin disorders can exacerbate into serious conditions when not tended to. At times a worker's skin disorder may be an allergic reaction which may not be solvable unless the worker is removed from that type of work. If the worker continues to do the same work this could result in a costly illness.

Physical agents, of which noise is the most common in the workplace, can lead to nonreparable hearing loss which becomes very compensable and degrades the value of that employee to you since he/she may not be able to hear warning signals or cannot communicate effectively with other workers. Although radiation (both ionizing and nonionizing) can be found in the workplace, it is not as common as noise, vibration, or temperature extremes.

What seems to present the most problems within the workplace are chemicals and the effects upon workers who are exposed to them. The major OSHA general standard which impacts most workplaces will be discussed. That standard is the Hazard Communication Standard.

2. Temperature Extremes

(1) Cold Stress

Temperature is measured in degrees Fahrenheit (°F) or Celsius (°C). Most people feel comfortable when the air temperature ranges from 66°F to 79°F and the relative humidity is about 45 percent. Under these circumstances, heat production inside the body equals the heat loss from the body, and the internal body temperature is kept around 98.6°F. For constant body temperature, even under changing environmental conditions, rates of heat gain and heat loss should balance. Heat loss is greatest if the body is in direct contact with cold water. Heat is also lost from the skin by contact with cold air. At a given air temperature, heat loss increases with air speed. Sweat production and its evaporation from the skin also cause heat loss. This is important when performing hard work.

The body maintains heat balance by reducing the amount of blood circulating through the skin and outer body parts. This minimizes cooling of the blood by shrinking the diameter of blood vessels. At extremely low temperatures, loss of blood flow to the extremities may cause an excessive drop in tissue temperature resulting in damage such as frostbite, and by shivering which increases the body's heat production. This provides a temporary tolerance for cold but cannot be maintained for long periods.

Overexposure to cold causes discomfort and a variety of health problems. Cold stress impairs performance of both manual and complex mental tasks. Sensitivity and dexterity of fingers lessen in cold. At lower temperatures still, cold affects deeper muscles, resulting in reduced muscular strength and stiffened joints. Mental alertness is reduced due to cold-related discomfort. For all these reasons accidents are more likely to occur in very cold working conditions.

The main cold injuries are frostnip, frostbite, immersion foot and trenchfoot, which occur in localized areas of the body. Frostnip is the mildest form of cold injury. It occurs when earlobes, noses, cheeks, fingers, or toes are exposed to cold. The skin of the affected area turns white. Frostnip can be prevented by warm clothing and is treated by simple rewarming.

Although people easily adapt to hot environments, they do not acclimatize well to cold. However, frequently exposed body parts can develop some degree of tolerance to cold. Blood flow in the hands, for example, is maintained in conditions that would cause extreme discomfort and loss of dexterity in unacclimatized persons. This is noticeable among fishermen who are able to work with bare hands in extremely cold weather.

Protective clothing is needed for work at or below 40°F. Clothing should be selected to suit the cold, level of activity, and job design. Clothing should be worn in multiple layers which provide better protection than a single thick garment. The layer of air between clothing provides better insulation than the clothing itself. In extremely cold conditions, where face protection is used, eye protection must be separated from respiratory channels (nose and mouth) to prevent exhaled moisture from fogging and frosting eye shields.

(2) **Heat Stress**

Operations involving high air temperatures, radiant heat sources, high humidity, direct physical contact with hot objects, or strenuous physical activities have a high potential for inducing heat stress in employees engaged in such operations. Such places include: iron and steel foundries, nonferrous foundries, brick-firing and ceramic plants, glass products facilities, rubber products factories, electrical utilities (particularly boiler rooms), bakeries, confectioneries, commercial kitchens, laundries, food canneries, chemical plants, mining sites, smelters, and steam tunnels. Outdoor operations, conducted in hot weather, such as construction, refining, asbestos removal, and hazardous waste site activities, especially those that require workers to wear semi-permeable or impermeable protective clothing, are also likely to cause heat stress among exposed workers.

Age, weight, degree of physical fitness, degree of acclimatization, metabolism, use of alcohol or drugs, and a variety of medical conditions, such as hypertension, all affect a person's sensitivity to heat. However, even the type of clothing worn must be considered. Prior heal injury predisposes an individual to additional injury. It is difficult to predict just who will be affected and when, because individual susceptibility varies. In addition, environmental factors include more than the ambient air temperature. Radiant heat, air movement, conduction, and relative humidity all affect an individual's response to heat.

The human body can adapt to heat exposure to some extent. This physiological adaptation is called acclimatization. After a period of acclimatization, the same activity will produce fewer cardiovascular demands. The worker will sweat more efficiently (causing better evaporative cooling), and thus will more easily be able to maintain normal body temperature. A properly designed and applied acclimatization program decreases the risk of heat-relate illnesses. Such a program basically involves exposing employees to work in a hot environment for progressively longer periods.

3. Ionizing Radiation

Ionizing radiation has always been a mystery to most people. Actually, much more known about ionizing radiation than the hazardous chemicals that constantly bombard the workplace. After all, there are only four types of radiation (alpha particles, beta particles, gamma rays, and neutrons) rather than thousands of chemicals. There are instruments that can detect each type of radiation and provide an accurate dose-received value. This is not so for chemicals, where the best that we could hope for in a real time situation is a detection of the presence of a chemical and not what the chemical is. With radiation detection instruments the boundaries of contamination can be detected and set, while detecting such boundaries for chemicals is near to impossible except for a solid.

It is possible to maintain a lifetime dose for individuals exposed to radiation. Most workers wear personal dosimetry, which provides reduced levels of exposure. The same impossible for chemicals where no standard unit of measurement exists for radioactive chemicals. The health effects of specific doses are well known such as 20–50 rems—when minor changes in blood occur, 60–120 rems—vomiting occurs but no long-term illness, or 5,000–10,000 rems—certain death within 48 hours. Certainly radiation can be dangerous, but one or a combination of three factors, distance, time, and/or shielding can usually used to control exposure. Certainly distance is the best since the amount of radiation from a source drops off quickly as a factor of the inverse square of the distance, for instance, at eight feet away the exposure is 1/64th of the radiation emanating from the source. As for time, many radiation workers are only allowed to stay in a radiation area for a certain length of time, and then they must leave that area. Shielding often conjures up lead plating or lead suits (similar to when x-rays are taken by a physician or dentist). Wearing a lead suit may seem appropriate but the weight alone can be prohibitive. Lead shielding can be used to protect workers from gamma rays (similar to x-rays). Once they are emitted, they could pass through anything in their path and continue on their way, unless a lead shield is thick enough to protect the worker.

New Words and Expressions

trauma ['trɔːmə]	n.	[医] 外伤，损伤
anthrax ['ænθræks]	n.	[兽] 炭疽热
chronic ['krɒnɪk]	adj.	慢性的，延续很长的
onset ['ɒnset]	n.	[医] 发作
latency ['leɪtənsi]	n.	潜伏，反应时间
asbestosis [ˌæzbes'təʊsɪs]	n.	[医] 石棉沉滞症
thyroid ['θaɪrɔɪd]	n.	甲状腺，甲状软骨
pancreatic [ˌpæŋkrɪ'ætɪk]	adj.	胰腺的
demise [dɪ'maɪz]	n.	死亡
elusive [ɪ'ljuːsɪv]	adj.	难懂的
rash [ræʃ]	n.	[医] 皮疹
nonreparable [ˌnaɪn'repərəbl]	adj.	不可恢复的，不可赔偿的
vessel ['vesl]	n.	脉管，导管
frostbite ['frɒstbaɪt]	n.	冻伤
dexterity [deks'terɪti]	n.	灵巧，机敏
stiffen ['stɪfn]	v.	使硬，使僵硬；变硬
joint [dʒɔɪnt]	n.	关节
immersion foot		[医] 足浸病
acclimatize [ə'klaɪmətaɪz]	v.	使适应（或习惯）新环境（或气候等）
insulation [ˌɪnsjʊ'leɪʃən]	n.	绝缘
respiratory [rɪs'paɪərətəri]	adj.	呼吸的
exhale [eks'heɪl]	v.	呼气
nonferrous ['nɒn'ferəs]	adj.	不含铁的，非铁的
ceramic [sɪ'ræmɪk]	adj.	陶器的
boiler room		锅炉房，锅炉间
semi-permeable ['semɪ'pɜːmɪəbl]	adj.	半渗透性的
acclimatization [əˌklaɪmətaɪ'zeɪʃən]	n.	环境适应性

metabolism [me'tæbəlɪzəm]	n. 新陈代谢；变形
hypertension [ˌhaɪpə'tenʃən]	n. 高血压；过度紧张
predispose ['priːdɪs'pəʊz]	v. 预先安排；使偏向于
ambient ['æmbɪənt]	adj. 周围的
cardiovascular [ˌkɑːdɪəʊ'væskjʊlə]	adj. 心脏血管的
progressively [prə'gresɪvli]	adv. 日益增多地
dosimetry [dəʊ'sɪmɪtri]	n. 放射量测定，剂量测定
vomit ['vɒmɪt]	v. 呕吐
shielding can	隔离罩
emanate ['eməneɪt]	v. 散发，发出；发源

Unit Seven

Safety Culture

1. The Concept of Corporate Culture

In response to the recognition that its structure has limitations in providing the "glue" that holds organisations together, much management thinking over the last two decades has focused on the concept of corporate culture. Some of the writings on the topic have been extremely influential among practising managers, mainly via its assumed relationship with organisational performance. It is generally thought that a well-developed and business-specifc culture into which managers and employees are thoroughly socialised will lead to stronger organisational commitment, more efficient performance and generally higher productivity. Usually based upon a blend of visionary ideas, corporate culture appears to reflect shared behaviours, beliefs, attitudes and values regarding organisational goals, functions and procedures which are seen to characterise particular organisations. The maintenance of the dominating corporate culture within any organisation, therefore, is supported by ongoing analyses of organisational systems, goal-directed behaviour, attitudes and performance outcomes. However, due to a general lack of information on how culture works, or how it can be shaped, changed or otherwise managed in practise, there is no consistent definition of what corporate culture might be. The main difference between such definitions appears to reside in their focus on the way people think, or on the way people behave, although some focus on both the way people think and behave.

Williams et al take issue with the notion that organisational culture reflects shared behaviours, beliefs, attitudes and values. They argue that not all organisational members respond in the same way in any given situation, although there may be a tendency for them to adopt similar styles of dress, modes of conduct, and perceptions of how the organisation does, or should, function. Beliefs, attitudes and values about the organisation, its function or purpose can vary from division to division, department to department, workgroup to workgroup, and from individual to individual. Thus, although an organisation may possess a dominating "cultural theme", there are likely to be a number of variations in the way in which the theme is expressed throughout the organisation. For example, one department may put safety before production, whereas another department may put production before safety. In the former, risk assessments might always be conducted prior to starting every job, while in the latter, people circumvent all the safety rules and procedures to ensure continuation of production. It follows, therefore, that several different sub-cultures will emerge from, or form around, functional groups, hierarchical levels and organisational roles, with very few behaviours, beliefs, attitudes or values being commonly shared by the whole of the organisation's membership. In turn, these sub-cultures may either be in alignment, or at odds, with the dominating "cultural" theme. This is not surprising given that organisations are dynamic, multi-faceted human systems that operate in dynamic environments in which what exactly suits at one time and one place cannot be generalised into a detailed universal truth" argues that differing sub-cultures actually serve a useful function, as they are a valuable resource for dealing with collective ignorance determined by systemic uncertainty because they provide a diversity of perspectives and interpretation of emerging (safety) problems.

2. The Concept of Safety Culture

The term "safety culture" first made its appearance in the 1987 OECD Nuclear Agency report on the 1986 Chernobyl disaster. Gaining international currency over the last decade, it is loosely used to describe the corporate atmosphere or culture in which safety is understood to be, and is accepted as, the number one priority. Unless safety is the dominating characteristic of corporate culture, which arguably it should be in high-risk industries, safety culture is a sub-component of corporate culture, which alludes to individual, job, and organisational features that affect and influence health and safety. As such the dominant corporate culture and the prevailing context such as downsizing and organisational restructuring will exert a considerable influence on its development and vice-versa as both inter-relate and reinforce each other. This latter point illustrates that safety culture does not operate in a vacuum: it affects, and in turn is affected by, other non-safety-related operational processes or organisational systems.

3. Definitions of Safety Culture

Numerous definitions of safety culture abound in the academic safety literature. Uttal, for example, defined it as "shared values and beliefs that interact with an organisation's structures and control systems to produce behavioural norms". Turner et al., defined it as, "the set of beliefs, norms, attitudes, roles, and social and technical practices that are concerned with minimising the exposure of employees, managers, customers and members of the public to conditions considered dangerous or injurious". The International Atomic Energy Authority defined safety culture as, "that assembly of characteristics and attitudes in organisations and individuals, which establishes that, as an overriding priority, nuclear plant safety issues receive the attention warranted by their significance". The Confederation of British Industry defined safety culture as, "the ideas and beliefs that all members of the organisation share about risk, accidents and ill health". The Advisory Committee for Safety in Nuclear Installations, subsequently adopted by the UK Health and Safety Commission (HSC), defined it as, ... the product of individual and group values, attitudes, competencies, and patterns of behaviour that determine the commitment to, and the style and proficiency of, an organisation's health & safety programmes. Organisations with a positive safety culture are characterised by communications founded on mutual trust, by shared perceptions of the importance of safety, and by confidence in the efficacy of preventative measures.

All these definitions are relatively similar in that they can be categorised into a normative beliefs perspective in so far as each is focused to varying degrees on the way people think and/or behave in relation to safety. Likewise, with the exception of the HSC these definitions tend to reflect the view that safety culture "is" rather than something that the organisation "has". In the former, safety culture is viewed as an emergent property of social groupings, reflecting an "interpretative view" favoured by academics and social scientists, whereas the latter reflects the functionalist view that culture has a pre-determined function favoured by managers and practitioners. It has been argued that both views are commensurate in that managerial functionalist strategies emerge from interpretative contexts. This appears to be the case with the HSC's definition, which takes the view that safety culture is a product emerging from values, attitudes, competencies, patterns of behaviour, etc. As such it reflects both a functionalist view of "culture" in terms of purpose and an interpretative view in that safety culture is also an emergent property created by social groupings within the workplace, indicating that normative beliefs are both created by, and revealed to, organisation members within a dynamic reciprocal relationship. Nevertheless, Cox suggest that the very broadness of the above definitions weaken their scientific utility, indicating that much greater precision is required.

Given the prominence of the HSC's definition in guiding researchers, one area requiring precision that appears to have been overlooked by all concerned is the "product" of the safety culture construct. This oversight has led to an overly narrow emphasis on safety climate via questionnaire surveys with it being used as a surrogate measure of safety culture, at the expense of the holistic, multi-faceted nature of the concept of safety culture itself. Defining this product is important as it could help to clarify what a safety culture should look like in an organisation. It could also help to determine the functional strategies required to develop the "product", and it could provide an outcome measure to assess the degree to which organisations might or might not possess a "good" safety culture. In turn, this could help to minimise the current unsystematic and fragmented approaches to researching safety culture and allow meta-analyses to be conducted at some time in the future. One conceptualisation that is consistent with the assessment characteristics of culture, with the fact that culture belongs to a group of people. In the current context, "effort" refers to the interaction between intensity and persistence of energy expenditure. In other words, how much energy a person expends to improve safety, and for how long in the face of obstacles. It is self-evident that what constitutes the units of "effort" could differ in different organisations. Nonetheless, the degree to which members consistently confront others about their unsafe acts, the degree to which members report unsafe conditions, the speed with which members implement remedial actions, the degree to which members give priority to safety over production are all observable examples of members directing their efforts to improve safety. Once these units of "effort" are identified, it is a relatively simple procedure to develop checklists with which to measure people against.

Although tentative, this definition of the safety culture "product" does at least provide an ongoing, tangible outcome measure that has been severely lacking, hitherto. Some might argue that reductions in accident rates provide a better outcome measure of safety culture. However, accident rates can be reduced for a number of reasons that have little to do with "safety culture" per se (e.g. under-reporting as a result of incentive schemes). Even if an organisation did actually achieve a genuine zero accident rate, this outcome measure would suffer from a lack of ongoing evaluative data, making it difficult, if not impossible, to determine the quality of its ongoing "safety culture". Thus, reductions in accident and injury rates, although very important, are not sufficient in themselves to indicate the presence or quality of a safety culture, whereas "that observable degree of effort…" is something that can always be measured and assessed.

Given that the maintenance of organisational cultures are supported by ongoing analyses of organisational systems, goal-directed behaviour, attitudes and performance outcomes, the definition given for the safety culture "product" provides a dependent variable with which to assess safety culture improvement initiatives. As such, it should become possible to M. D. Cooper empirically examine the links between those personal (e.g. values, beliefs, attitudes), behavioural (e.g. competencies, patterns of behaviour) and situational (e.g. organisational systems and sub-systems) aspects of safety culture reflected in the above definitions, to determine their impact on the development of its "product". Again, these links could and should be examined singly and in combination, at the level of the individual, the workgroup and the organisation. In this way, the most effective aspects for developing the safety culture "product" can be identified, which in turn may help to redefine the concept itself with much greater precision.

4. Strengthening the Concept of Safety Culture

Despite notions that culture cannot easily be created or engineered, in practise, the creation or enhancement of a safety culture is dependent upon the deliberate manipulation of various organisational characteristics thought to impact upon safety management practices. The very act of doing so means that the manipulations must be goal-directed. Because goals (ideas of future, ideas of a desired end-state) play a strong causal role in action, Locke and Latham's Goal-Setting Theory may also serve to provide the requisite scientific utility sought by Cox. This becomes apparent when the specific purposes of the safety culture definitions outlined above are examined. These include: ①producing behavioural norms; ②reductions in accidents and injuries; ③ensuring that safety issues receive the attention warranted by their significance; ④ensuring that organisational members share the same ideas and beliefs about risks, accidents and ill-health; ⑤increasing people's commitment to safety; and ⑥determining the style and proficiency of an organisation's health and safety programmes. Each of these purposes can be viewed both as sub-goals (i.e. antecedents) that help an organisation to attain its superordinate goal, and goal-achievements emanating from the creation of an organisation's safety culture.

If these Goal Theory concepts are accepted, the creation of a safety culture simply becomes a super-ordinate goal, that is achieved by dividing the task into a series of sub-goals that are intended to direct people's attention and actions towards the management of safety. In goal-theoretic terms, performance is a positive function of goal-difficulty. The greater the challenge, the better people's performance tends to be. Setting a difficult super-ordinate goal will therefore place challenging demands on individuals, workgroups, departments, and the organisation as a whole. Dividing the task into more manageable sub-goals that are in themselves challenging and difficulty should lead to much greater overall goal-attainment of the super-ordinate goal. Nonetheless, goal-attainment is known to be affected by a number of moderators such as ability, goal-commitment, goal-conflict, feedback, task complexity; and, situational constraints, as well as mediators such as direction of attention, effort and persistence, task-specific strategies and self-efficacy.

In safety culture terms these goal-related moderators could be viewed as being analogous to safety- and job-related training; degrees of commitment to safety at various hierarchical levels; safety versus productivity, quality; communication flows; managerial versus operative's role functions; and lack of resources, work-pace, job design issues. Similarly, the goal-related mediators could also be translated into safety culture terms. For example, direction of attention, effort, and persistence could reflect actual safety-related behaviour (s) at different hierarchical levels of the organisation. The presence and quality of the organisation's decision-making processes could be analogous to task-specific strategies. Self-efficacy could be translated into individuals', workgroups', departments', and/or business units' confidence in pursuing particular courses of action to bring about safety improvements. Importantly, each of these moderators and mediators can be examined individually and in various combinations to assess their impact on both the achievement of sub-goals and the super-ordinate goal.

New Words and Expressions

circumvent [ˌsɜːkəmˈvent]	v. 围绕
hierarchical [ˌhaɪəˈrɑːkɪkəl]	adj. 分等级的
alignment [əˈlaɪnmənt]	n. 排列成行, 队列; 结盟
allude [əˈljuːd]	v. 暗指, 影射, 间接提到
warrant [ˈwɒrənt]	v. 批准; 使有正当理由
normative [ˈnɔːmətɪv]	adj. 标准化的
functionalist [ˈfʌŋkʃənəlɪst]	n. 机能主义者
commensurate [kəˈmenʃərɪt]	adj. 相称的, 相当的
reciprocal relationship	互反关系
prominence [ˈprɒmɪnəns]	n. 突出, 显著, 突出物
oversight [ˈəʊvəsaɪt]	n. 勘漏, 失察, 疏忽
surrogate [ˈsʌrəgɪt]	n. 代理
holistic [həʊˈlɪstɪk]	adj. 整体的, 全盘的
remedial [rɪˈmiːdjəl]	adj. 补救的; 治疗的
tentative [ˈtentətɪv]	adj. 试验性的, 试探的, 尝试的; 暂定的
tangible [ˈtændʒəbl]	adj. 切实的
hitherto [ˌhɪðəˈtuː]	adv. 迄今, 至今
dependent variable	因变量, 应变数
empirically [emˈpɪrɪkəli]	adv. 以经验为主地
antecedent [ˌæntɪˈsiːdənt]	n. 先辈
superordinate [ˌsjuːpəˈɔːdɪnɪt, -ˌneɪt]	adj. 高级的; 地位高的, 上级的
moderator [ˈmɒdəreɪtə]	n. 缓和剂, 慢化剂
analogous [əˈnæləgəs]	adj. 类似的, 可比拟的

Translation Skill

科技英语翻译技巧（六）——常见多功能词的译法（II）

一、that 的译法

1. that 作为限定词, 可译成"那, 该"等

Up to *that* time wood and stone were still the main building materials.

直到那时，木材和石头仍是主要建筑材料。

2. that 作为指示代词的译法

1）that 代表前面的句子时，可译成"这"。

Plastics are light and strong and do not rust at all. *That* is why they find such wide uses in industry.

塑料既轻又牢，毫不生锈。这就是塑料在工业上如此广泛应用的原因。

2）that 代替前面的某一名词时，翻译时往往重复所代替的名词。

Mild steel has a high tensile strength, this being 200 times *that* of concrete.

低碳钢具有很高的抗拉强度，大约是混凝土的抗拉强度的 200 倍。

3. that 作为关系代词的译法

that 作为关系代词，引出定语从句；定语从句可以合译也可分译，that 的处理方法也随之而异。

1）合译时 that 不译，在定语从句之末添"的"字。

Power is the rate *that* mechanical move is performed at.

功率是作机械运动的速率。

2）分译时，that 有两种译法：a）重复所代替的名词；b）译成"它"。

Matter is composed of molecules *that* are composed of atoms.

物质由分子组成，而分子由原子组成。

An element is a simple substance *that* cannot be broken up into anything simpler.

元素是一种单质，它不能再分成任何更简单的物质。

4. that 作为连词引出同位语从句的译法

1）同位语从句与主句分译，that 译成"即"。

Energy takes many forms, but all these forms can be reduced to the statement *that* energy is the capacity to do work.

能量具有很多形式，但所有这些形式都可以归纳为这样一句话，即能量是做功的能力。

2）同位语从句与主句合译，that 译成"这一"。

The idea *that* energy is conserved is the first law of thermodynamics.

能量守恒这一概念是热力学第一定律。

3）在主句与同位语从句之间加冒号，that 略去不译。

In 1905 Einstein worked out a theory *that* matter and energy were not completely different things.

1905 年，爱因斯坦提出了一个理论：物质和能量并非完全不同的东西。

5. that 引出表语从句时，一般略去不译，如表语从句较长，可在主句之间增加冒号，或增加"在于"

One of the advantages of concrete is *that* it can be easily shaped into any desired form.

混凝土的优点之一是它易于制成所需的任何形状。（略去不译）

The law of electric current is *that* an electric current varies directly as the voltage and inversely proportional to the resistance.

电流定律就是：电流与电压成正比，与电阻成反比。

6. that 引出主语从句、宾语从句；在强调句型时，一律略去不译；有时在宾语从句前加上逗号

It is generally believed *that* oil is derived from marine plant and animal life.

通常认为石油来源于海生动植物。（引出主语从句）

7. that 作为连词引出结果状语从句和目的状语从句；前者译成"因而，以致，从而"等，后

者可译成"才,以便"等

The climate is very hot and dry *that* much evaporation takes place.
气候既很炎热又很干燥,因而产生大量的蒸发。
The parts are of enough strength *that* they may not break in use.
零件具有足够的强度,才不会在使用中破损。

二、what 的译法

1. what 作为连接代词可译为"所谓,什么……"或可根据上下文译成名词

What is large and *what* is small is only relative.
所谓大和小只是相对的。
Hydrogen and oxygen are *what* make up water.
氢和氧是组成水的元素。

2. what 作为关系代词,可译为"所……的东西,所……的",还可根据上下文译成"……的+名词,那种+名词"等

What is worrying the world greatly now is a possible shortage of coal, oil, natural gas, or other sources of fuel in the not too distant future.
现在令全世界所担心的是,在不久的将来可能会出现煤、石油、天然气或其他能源的短缺。

3. what 作为限定词,此时 what 在从句中做定语,可译成"什么,什么样的,哪"等

One of the important problems to be solved is *what* material is most suitable for this particular part.
需要解决的重要问题之一是,对于这个特定的零件用什么材料最合适。

Reading Material

Perspectives on Safety Culture

1. Organisational Culture and Safety Culture

Organisational culture, however defined, is widely acknowledged to be critical to an organisation's success or failure, for example in business. Graves and Williams et al. consider that the prime function of culture is to contribute to an organisation's success. Analogously, safety culture is frequently identified, for example by disaster inquiries, as being fundamental to an organisation's ability to manage safety-related aspects of its operations—successfully or otherwise. Implicit within both these views is the notion that culture operates at different levels and through various mechanisms. However, the nature of these mechanisms remains problematic.

Because the notion of safety culture arose from the more inclusive concept of organisational culture, some key features characterising debate about this concept are first considered. Broader issues, including derivation of the notion of culture from social, ethnic or other origins, are excluded here.

Waring and Glendon review approaches to organisational culture from two contrasting perspectives that have dominated the literature, as well as managerial and professional practice within organisations. These two broad perspectives have been described as functionalist and interpretive. Waring considers that functionalist approaches assume that organisational culture exists as an ideal to which organisations should aspire so that it can, and should be, manipulated to serve corporate interests. The notion that organisational culture has, as its prime function, to support management strategies and systems is premised on the assumption that it can be reduced to relatively simple models of prediction and control. This approach primarily aligns organisational culture in support of managerial ideology, goals and strategy, in extreme cases involving managerial use of "culture" to coerce and control. Ideological use of culture as a weapon in organisational struggles reveals a powerful bias. This engineering model of organisational culture is criticised by Sackmann, who regards it as problematic as to whether leaders initiate culture.

Alternative expositions of organisational culture can provide a more complete understanding of this important concept. From this imperative derived interpretive approaches to organisational culture. Interpretive approaches assume that organisational culture is an emergent complex phenomenon of social groupings, serving as the prime medium for all members of an organisation to interpret their collective identity, beliefs and behaviours. Organisational culture is not owned by any group but is created by all the organisation's members. Consonant with an interpretive perspective is Schein's developmental approach to organisational culture, defined as a pattern of assumptions developed by a group as it learns to adapt to its environment. The culture is taught to new members as the framework for cognitions and behaviours in response to problems.

From assumptions characterising interpretive approaches to organisational culture, it follows that managerial attempts to manipulate culture, for example in seeking to drive rapid organisational change, are likely to fail because of the application of an inadequate model of processes that they attempt to manipulate. An analogous point, in respect of organisations seeking to enhance safety culture as the philosopher's stone to improving health and safety, is made by Cox and Flin.

A functionalist perspective is "top down" in that it serves the strategic imperative of members of the controlling group. An interpretive perspective represents a "bottom up" approach, and allows for the existence of sub-cultures within an organisation. Most organisations display elements of both approaches. For example, through rigorous adoption of formalised risk management practices, an organisation reveals a functionalist approach to culture. A more interpretive side may be revealed by individual and group commitment to learning from past mistakes, such as those leading to accidents. This might be achieved through open-ended communication and discovery processes, involving a developing identity for the organisation's members.

As Waring and Glendon observe, from an interpretive standpoint, culture provides a metaphor for understanding how organisations work and why they respond in particular ways to environmental influences. These authors argue that an interpretive perspective on culture is more appropriate than a functional approach as a way of modeling attempts to understand behaviours and cognitions in respect of safety and other aspects of organisational life.

A number of classifications have been suggested for organisational culture. A global approach within one organisation was the basis for Hofstede's well-known taxonomy—a tradition continued through the current 65-nation GLOBE project. Furnham and Gunter's culture taxonomy is based on theoretical versus empirical origin. The former they identify as being top down approaches that are based on conceptual distinctions from previous work. Empirical approaches are identified as being bottom up and data-driven to produce a set of dimensions for defining culture, but probably theoretically void. Considerable emphasis in the literature has been upon seeking appropriate measures, dimensions and taxonomies for organisational culture, in part at least to find a way to an optimum culture. However, this is premised upon a functionalist approach as being the best means of understanding culture.

2. Dimensions of Organisational Culture

A number of attempts have been made to map the main features or levels of organisational culture. A large degree of concurrence exists between espoused models. For the content of organisational culture, a three-level classification embodying relatively accessible, intermediate and deep levels, forms the basis for the composite model outlined. The most accessible level refers to observable behaviours and perhaps associated norms. At an intermediate level are attitudes and perceptions, which are not directly observable, but which may either be inferred from behaviours or assessed through questioning. At the deepest level are core values, which are much less amenable to assessment and for whose investigation more ethnographic methods are likely to be required.

Other key dimensions of organisational culture that have been identified include depth, breadth and progression. Depth refers to the way in which culture is reflected in the organisation's policies, programs, procedures, practices, values, strategies, behaviours and other features. Cultural breadth is represented in the lateral coordination of different organisational components. Progression refers to the time dimension, and is similar to the developmental aspect of culture espoused by Schein. Gorman identifies three further dimensions. Strength is the extent to which organisation members embrace core level meanings. Pervasiveness refers to the extent to which beliefs and values are shared across the organisation. Direction reflects the extent to which organisational culture embodies behaviour that is consistent with espoused strategy. Waring and Glendon, following Turner's pluralistic notion of organisations as assemblages of multiple cultures, add localisation, this being the extent to which organisational locations exhibit sub-cultures. Schein identifies seven dimensions of organisational culture, and considers that critical dimensions of culture—reflecting its strength and degree of internal consistency —are defined by the stability of a group and how long it has existed. Also important is the intensity of group learning experiences, how learning occurred, and the strength and clarity of assumptions held by group founders or leaders. If organisational culture, or some aspect of it, is to be measured at three levels and across several dimensions, then complex and imaginative methods of assessment and analysis will be required. Questionnaire or similar measures will be inadequate to measure all aspects of organisational culture. Validated questionnaires are acceptable as climate measures. Organisational climate and its derivatives might be comparable with intermediate levels of culture measured across some of the dimensions already outlined. This issue is explored further later.

Rousseau reviews several instruments supposedly designed to measure organisational culture, and found considerable variation in what was measured and in the extent to which validity and other methodological issues were addressed. Broadfoot and Ashkanasy reviewed 18 survey instruments designed to measure organisational culture, all of which they found exhibited serious flaws. Contemporary circumstances, within which measurement of all aspects of managerial performance becomes an imperative, means that pressure to assess culture and its derivatives can be difficult to resist. However, exclusively functionalist approaches to the measurement of organisational culture are likely to be inadequate because they are based upon an incomplete model of the concept for which measurement is sought.

To comprehensively assess organisational culture, or some aspect of it—such as safety culture—the measures used must be based upon an adequate model of culture, taking account of its multi-faceted nature. Three methodologies that have been used to assess and analyse organisational culture are:
- Soft systems methodology adopts a broad perspective that can incorporate both quantitative and qualitative data.
- Organisational climate surveys supported by triangulated methods.
- Grid-group analysis.

Locatelli and West examined three qualitative approaches to measuring organisational culture—repertory grids, the twenty statements test (TST) and group discussions. The criteria used by Locatelli and West to compare the three methods were:
- Level of cultural information accessed, including specific cultural elements (artefacts, values, underlying beliefs about which information was elicited).
- Quality and usefulness of information generated.
- Ease of use, including time and cost.

Locatelli and West showed that TST performed best, being both quickest and producing the most relevant information. However, there was poor overall inter-rater agreement and no one method had comprehensive coverage for all aspects of the organisational culture framework. A combination of methods is indicated, probably comprising as a minimum the TST and a questionnaire—assuming that together these will capture information on values and beliefs as well as underlying assumptions. A more comprehensive analysis might also require checklists (to capture data on artefacts) and activity analysis (for information on behaviour patterns). Although no study has yet reported on the combination of all these approaches in the open literature, a case study combining questionnaires and human factors interventions is discussed later.

3. Organisational Culture and Organisational Climate

Confusion between the terms "culture" and "climate" means that they have been used interchangeably. While there is a relationship and some overlap between these terms, organisational climate refers to the perceived quality of an organisation's internal environment. In a review by Rousseau of 13 definitions derived over a 21-year period, employee attitudes and perceptions featured prominently. Typical was the definition proffered by James and Jones, of psychologically meaningful cognitive representation of the situation. Furnham and Gunter regard organisational climate as being an index of organisational health, but not a causative factor in it.

Typically organisational climate is regarded as a more superficial concept than organisational culture, describing aspects of an organisation's current state. Scaled dimensional measures are the most popular means of measuring organisational climate, of which many have been devised. There is no agreement on the key dimensions to be measured. Furnham and Gunter identify 35 possible scales, one of which is risk. In reviewing the literature, Koys and De Cotiis produced a composite eight-dimensional scale with the components. These generic categories essentially relate to human resource aspects of an organisation's environment, and while substantive elements of organisational life, such as safety or risk do not figure, dimensions from Koys and De Cotiis typically feature on safety climate scales.

Methodology might be a good indicator as to whether organisational culture or climate is being measured. If a psychometric scale is the exclusive measurement instrument, then some aspect of organisational climate is being measured. A triangulated methodology might indicate that other aspects of culture were being tapped, although this would depend upon the depth and breadth of the measures used. What organisational climate measures may access are some dimensions of organisational culture within a limited range. For example, climate questionnaires might access attitudes, beliefs and perceptions that are located at the mid-range of cultural levels. However, surveys are limited by their methodology and can only report on attitudes at the time that they are undertaken and perhaps also a little in the past. Thus, organisational climate surveys might provide a snapshot of selected aspects of organisational culture. Without validation, climate survey findings may be difficult to interpret within a culture framework.

By imposing a unified mono-culture, a functionalist approach to organizational (or safety) culture or climate renders the notion of sub-cultures largely redundant. Aggregating questionnaire scores across departments or other groupings within organisations predicate a functionalist approach. Acknowledging diversity from such surveys would go some way towards identifying a bottom-up approach to these phenomena. However, the methodology imposes uniformity upon the data. Identifying sub-cultures and the potential diversity that this implies as a basis for improving understanding of the phenomena under study could be a valuable way forward.

(1) **From Organisational Culture to Safety Culture**

Contrasting perspectives on organisational culture can be used as a framework for appreciating how values, attitudes and beliefs about safety are expressed and how they might influence directions that organisations take in respect of safety culture. The term "safety culture" arose from the Chernobyl nuclear disaster in 1986, in which cause was attributed to a breakdown in the organisation's safety culture. Subsequently, the concept was heralded as a substantive issue in official inquiry reports into disasters such as Kings Cross and Piper Alpha. The term rapidly gained currency within the safety management lexicon.

As with the concept of organisational culture, a range of meanings has been attached to safety culture, three of which are reviewed by the Institution of Occupational Safety and Health. The first meaning includes those aspects of culture that affect safety. The second refers to shared attitudes, values, beliefs and practices concerning safety and the necessity for effective controls. The third relates to the product of individual and group values, attitudes, competencies and patterns of behaviour that determine the commitment to, and the style and proficiency of, an organisation's safety programs. The latter two definitions are premised upon a functionalist perspective. Booth proposes an audit approach, while Cooper's methodology for changing safety culture incorporates risk assessments, audits, training, climate surveys and behaviour change. The approach taken by these authors implies that safety culture is conflict-free and aligned with the objectives of a controlling function.

(2) **Safety Climate**

Contemporaneous with the derivation of safety culture from organisational culture was the associated term "safety climate", which came from a more empirical tradition. Some researchers distinguish between safety culture and safety climate, while attempts have also been made to derive composite models. The prime research method for investigating safety climate is the questionnaire, typically completed by sufficient numbers of employees to enable statistical analysis to reduce a large number of items to a small number of dimensions. These dimensions are intended to represent the essence of safety climate for the organisation.

This empirical tradition has elements of both functionalist and interpretive perspectives. The methodology presumes that, much as in trait-based approaches to personality, organisations have safety climates that are waiting to be discovered. The measurement and inference that, once revealed the perceptions that comprise safety climate dimensions will be associated with measurable safety behaviours, with the implied targeting by management of these perceptions and behaviours, suggests a functionalist approach. However, the notion that safety climate derives essentially from aggregate employee perceptions, that it is multi-dimensional and that it can potentially influence safety-related behaviours, means that the concept belongs more in the interpretive school. Notwithstanding this conceptual position, finding an empirical association between safety climate dimensions and safety behaviour measures has so far proved elusive, although this could be due to methodological and analytical difficulties as much as to the presence or absence of such an association.

New Words and Expressions

aspire [əˈspaɪə]   v. 渴望，立志
ideology [ˌaɪdɪˈɒlədʒi]   n. 意识形态
coerce [kəʊˈɜːs]   v. 强制，强迫
bias [ˈbaɪəs]   n. 偏见，偏爱，偏差
rigorous [ˈrɪɡərəs]   adj. 严格的，严厉的，严酷的，严峻的
metaphor [ˈmetəfə]   n. 隐喻，暗喻
taxonomy [tækˈsɒnəmi]   n. 分类法，分类学
optimum [ˈɒptɪməm]   adj. 最适宜的
amenable [əˈmiːnəbl]   adj. 应服从的，有服从义务的，有责任的
lateral [ˈlætərəl]   adj. 横（向）的，侧面的
pervasiveness [pəˈveɪsɪvəs]   n. 无处不在，遍布
triangulate [traɪˈæŋɡjʊleɪt]   v. 分成三角形，对……作三角测量
elicit [ɪˈlɪsɪt]   v. 得出，引出
artefact [ˈɑːtɪfækt]   n. 人工制品，手工艺品
redundant [rɪˈdʌndənt]   adj. 多余的，冗余的
herald [ˈherəld]   v. 宣布，传达
lexicon [ˈleksɪkən]   n. 专门词汇
multi-dimensional [ˌmʌltɪdɪˈmenʃənl]   adj. 多面的，多维的

Unit Eight

Motivating Safety and Health

1. Defining Motivation

Some people believe motivation has the potential to answer everyone's problems. You may have heard such statements as "If I were only motivated!" "You should motivate me!" "You should motivate him or her!" "You are not motivated!" "You had better get motivated!" or "All you have to do is find his/her 'hot button'!" These statements, however, do not tell us what motivation really is nor do they tell us how we can measure or even understand motivation.

Motivation, in the broadest sense, is self-motivation, complex, and either need- or value-driven. Someone once stated that he believed "hope" was the secret ingredient to a person being motivated (the hope to accomplish a goal, a dream, or attain a need); there is reason to support this theory. But, possibly a better definition is "motivation presumes valuing, and values are learned behaviors; thus, motivation, at least in part, is learned and can be taught". This definition provides us with the encouragement we need in order to go forward and achieve motivation for ourselves and others.

If we want to be successful, we must believe that we can teach someone to be motivated toward specific outcomes (goals) or, at the least, be able to alter some unwanted behavior. On the other hand, we should not want to completely manipulate an individual to the point that he responds blindly to our motivational efforts.

Thus, motivation is internal. We cannot directly observe or measure it but a glimpse of the results may be observed when we see a positive change take place in behavior. Such a change might be something as simple as a worker wearing protective eyewear or something as far reaching as going a full year without an accident or injury. By observing these outwardly manifested behaviors, you can then be encouraged when you see even the smallest of successes that are related to your motivational techniques.

2. Principles of Motivation

Goals are an integral part of the motivational process and tend to structure the environment in which motivation takes place. The environment in which we find ourselves is many times the springboard to the overall motivational process. You may be fortunate enough to accidentally step into a high-energy motivational environment. On the other hand, you may find yourself in an environment that is not at all conducive to motivating others and, thereby, it is very difficult to attain your desired goals. If this is the case, you may need to make a change in the physical environment or possibly even make a change in the work atmosphere.

If you are that person with the responsibility of trying to motivate an individual or group, you will need to address their motivational needs. Employees fall along a continuum—some need little motivation from you and others need constant attention. It is unrealistic to expect all of them to achieve your level of expectations. The quality of your leadership will be the determining factor to your success with these people.

Why are some leaders more motivational than others? What are the unique talents which these dynamic leaders possess. Some people believe that these individuals were born to be leaders. Most of us do not believe that they are just "born leaders" but individuals who possess a set of talents and have chosen to develop those talents to the maximum. These talents are developed because they have the burning desire (goal) to become leaders and those desires motivate them to learn the necessary skills.

To be a successful motivational leader you must have some sort of plan that will get you from point "A" to point "B". This plan should include your desired goals and objectives, levels of expectations, mechanisms for communication, valuative procedures and techniques for reinforcement, feedback, rewards, and incentives. Any motivational plan is a dynamic tool that must be flexible enough to address changes, which may occur over a period of time and take into consideration the universality of people and situations. These plans can use a variety of techniques and gadgetry to facilitate the final desired outcomes or performances, which lead to a safer and healthier workplace.

3. Self-Motivation

The question that arises is, "Who motivates you?" Is it a person, is it peer groups, is it incentives, or is it the environment? It is the contention of this unit that no matter what, it is you who motivates you. Excuses, blame, and alibiing will not negate this fact. Nobody can motivate you. You must assume the responsibility to motivate yourself within the environment in which you find yourself. Some individuals are motivated by positive happenings within their lives, while others succeed through adversity. Certainly an employer may work very hard to set up a motivational environment, but the individual decides if he will be motivated by that environment.

One person who was motivated by his failures was Dan Jensen, Olympic speed skater. After failing to receive the gold medal in three previous Olympic Games, he went on to become a gold medal winner in the 1994 Winter Olympics Games. He had been expected to win the gold medal in previous Olympic Games but through disastrous falls or unexpected losses he was unable to accomplish that goal. He was determined to make his failure lead him to his success and ultimate goal, a gold medal in the 1994 Winter Olympics' 1000 meter race. Failures can bring success!

On the other hand, what would have happened to Bonnie Blair if she had experienced the same fate as Dan Jensen? She culminated her career with five gold medals and had at least one medal in each of the previous three Olympics. No one can say, or even guess the answer since she was motivated by her successes each time, instead of her failures. Mistakes can either have a positive or negative effect on the motivational environment, but tend to be de-motivational; we need to realize that people who are doing something are going to make mistakes.

Regardless of all the efforts made to assure that no accidents, injuries, or work-related illnesses occur, there will still be, at times, mistakes made and negative outcomes that occur. When it happens, this should be an incentive to try even harder; don't trash the safety and health effort over a setback.

The basis for motivation seems to be in our perceptions of ourselves. These perceptions govern our behavior and support the concept of self-motivation. In order for people to motivate themselves, there must be meaning in what they are doing. If they do not perceive that the goal set before them will satisfy their needs, they cannot possibly motivate themselves to accomplish it. You must realize that no matter how unrealistic a perception may seem to us, it is a reality to the person who holds it. No matter how we try to debunk a perception, there is always some degree of truth and reality within it and, therefore, it is very real to that person.

Individuals will not be motivated to work safely unless they have internalized the goals and expectations of the company. It is not enough for them to know that they will be fired for violation of a safety rule, they need to be motivated to perform their work safely even when there is no one watching them.

People must be inspired to be accountable to themselves. If they put their goals and plans down on paper, then they take possession of their own behavior to a greater extent. This motivates them to do something and gives them the time, direction, and a reason to find new or better ways to accomplish their goals and plans. As many experts will tell you, you should put your goals or plans in writing. If you can't write it down, then you probably will never achieve it.

The most important things to remember about people are: People are different with this in mind, you will need to view each person on a continuum. When trying to figure out what motivates him or her and how you can begin to get a change, take a look at every aspect concerning that person's life and try to evaluate what is and is not of importance to him or her. Some individuals are superstars. These individuals are self-motivated and all you have to do is give them support and minimal guidance and then just step back and watch them go. On the other hand, seem to lack any motivation at all. These individuals need to have things structured for them, know exactly what is expected of them, know what happens if they do not perform, and know what the reward or outcome of their performance will be.

You will also find individuals who need to be around other people; they perform best when they are in a social environment and, therefore, are more affected by peer group pressures. And, finally, there are the people who prefer to work alone. Frequently these individuals are achievement-oriented and all they want is your recognition and reinforcement. All it may take to motivate them is to grant their request for something as simple as a tool or piece of equipment, which will help them do their job in a safer manner.

4. Needs Move Mountains and People

Dr. Maslow identified five needs. They run the gamut from the basic animal need to the highly intellectual needs of modern man. They are the physiological, safety, social, and self-fulfillment needs. In order for you to understand the relationship between these needs and the motivational process, a description of each one follows.

Physiological needs are the requirements we have for our survival. They encompass the basic needs that are necessary for the body to sustain life or physical well-being. The needs are such things as the food we eat, the clothing we wear, and the shelter we live in. Each of these must be satisfied before other needs can be dealt with. The physiological needs appear in all of the actions each of us take to insure our survival and physical well-being. Individuals who are motivated primarily by these needs will do anything that you ask them to do no matter how unsafe it might be.

Safety needs include the requirements for our security. If first the physiological needs are reasonably well satisfied, then people become aware of and start to act to satisfy their safety needs. These needs are such things as having freedom from fear, anxiety, threat danger, and violence and being able to have stability in their lives. Striving to satisfy these needs might show up in such actions as: (for their safety) avoiding people or situations which are threat.

Social needs include the requirements for feeling loved and wanted, and the sense of belonging and being cared for. If the safety needs can be reasonably satisfied, social needs begin to emerge. Some behaviors that take place and indicate a social need of acceptance are asking the opinion of the group before acting, following group preferences instead of personal preferences, or, joining job-related interest groups. These individuals will follow the safety and health pattern set by their workgroup.

Ego needs include the requirements for self-identity, self-worth, status, and recognition. When social needs are reasonably satisfied, individuals are able to explore the dimensions of who they are and consider how they wish to sale/market themselves. Some examples of ego (esteem) needs are: self-respect, esteem of others, self-confidence, mastery competence, independence, freedom, reputation, prestige, status, fame, glory, dominance, attention importance, dignity, and appreciation. These individuals will want to be involved in and a part of the ongoing safety and health effort set by the company.

Self-fulfillment needs are composed of the requirements it takes to become all that one is capable of becoming and to fulfill oneself as completely as possible. The self-fulfillment needs are so complex that people never reach a point where they are completely fulfilled. They do what they do not because they want others to notice them or to reward them, but because they feel a need to be creative, to grow, to achieve, and to be all that they are capable of becoming. These individuals understand the true importance of safety and health on the job; it is a part of them. They will follow the safety and health rules because they have internalized the true function of the safety and health program; they realize that it is a vital component of the whole operation. These individuals have a sense of needing to help others reach an understanding of the safety and health issue.

Maslow was right when he suggested that needs are motivators for people. As a motivator, you cannot motivate another person by depending upon elements which you deem as important. What you must do is that be sensitive to the needs and wants of the people you are trying to motivate. It is sometimes difficult for many of us to remember where we came from and to relate to someone who has basic needs (physical and security) which are way below our own needs. If you are to be a real motivator, you will need to spend time understanding the real needs of those around you.

In summary, this means that each individual you are trying to motivate will need individualized attention. You will need to tailor, as best you can, a motivational plan which will meet his needs and, thus, causes him to be motivated toward the goals which have been developed. A person has his own reasons, based on his own values, needs, and desires which determine how they apply his own energies. How you accomplish this is not scientific. It may be accomplished by trial and error or, at best, by small successes followed by bigger successes until the goal is reached. It seems safe to say that what works well for one person may fail miserably for another or, with modification, may be successful.

New Words and Expressions

continuum [kən'tɪnjʊəm]	n.	连续统一体，连续统一
gadgetry ['gædʒɪtri]	n.	小配件
facilitate [fə'sɪlɪteɪt]	v.	推动，促进；使容易
contention [kən'tenʃən]	n.	争辩；论点
alibi ['ælɪbaɪ]	v.	辩解，找托词开脱
negate [nɪ'geɪt]	v.	否定，打消
culminate ['kʌlmɪneɪt]	v.	达到顶点
trash [træʃ]	v.	贬低，折掉，废弃
debunk [diː'bʌŋk]	v.	揭穿，拆穿假面具，暴露
violation [ˌvaɪə'leɪʃən]	n.	违反，违背
self-motivated ['selfməʊtɪvɪtɪd]	adj.	自我激励的
esteem [ɪs'tiːm]	n.	尊敬，尊重

Translation Skill

科技英语翻译技巧（七）——数词的译法

科技英语中大量出现表示数量、倍数增减的词语，在翻译时要特别注意英汉两种语言表达方式的异同。

一、as...as... 与数词

as + 形容词 + as + 数词 + ……
动词 + as + 副词 + 数词 + …… } 结构的译法

as + 形容词 + as + 数词 + ……

as large as + 数词……	可译成"大到（至）……"
as many as + 数词……	可译成"多达……"
as high as + 数词……	可译成"高达……"
as heavy as + 数词……	可译成"重达……"
as low as + 数词……	可译成"低到（至）……"

The temperature at the sun's center is *as high as* 10,000,000℃.
太阳中心的温度高达 1.0×10^7 ℃。

二、倍数的比较

A is N times as large (long, heavy...) as B
A is N times larger (longer, heavier...) than B } 结构的译法
A is larger (longer, heavier...) than B by N times

上述三种句型，虽然结构各异，但是所表达的概念完全一样。因此在译成汉语时彼此之间不应有所

区别。均可译成：A 的大小（长度、重量……）是 B 的 N 倍。或 A 比 B 大（长、重……）N－1 倍。

This substance reacts *three times as fast as* the other one.
这一物质的反应速率比另一物质快两倍。

In case of electronic scanning the beam width is *broader by a factor of two*.
电子扫描时，波束宽度扩大一倍。

三、倍数的增加

1) 主语 + 谓语 + { double / treble / quadruple } + …… 结构的译法。

double　　　可译成"增加一倍"或"翻一番"
treble　　　可译成"增加二倍"或"增加到三倍"
quadruple　　可译成"增加三倍"或"翻两番"

Whenever we *double* the force on a given object, we *double* the acceleration.
每当作用在某一给定物体上的力增加一倍，加速度也增加一倍。

2) { increase N times / increase to N times / increase by N times / increase N-fold / increase by a factor of N } 结构的译法。

上述 5 种结构表达的意思完全一致。均表示"乘以 N""成 N 倍"这一概念，因此应译成"增加到 N 倍"或"增加了 N－1 倍"。

A temperature rise of 100℃ increases the conductivity of a semiconductor *by 50 times*.
温度上升 100℃，半导体的电导率就增加到 50 倍。

Since the first transatlantic telephone cable was laid the annual total of telephone calls between UK and Canada has *increased sevenfold*.
自从第一条横跨大西洋的电缆敷设以来，英国与加拿大之间的年通话量增加了 6 倍。

3) { as much (many…) again as / again as much (many…) as } 可译为"是……的两倍"或"比……多一倍"
   { half as much (many…) again as / half again as much (many…) as } 可译成"是……的一倍半""比……多半倍"或"比……多一半"。

The amount left was estimated to be *again as much as* all the zinc that has been mined.
当时估计，剩余的锌储量是已开采量的两倍。

The resistance of aluminum is approximately *half again as great as* that of copper for the same dimensions.
尺寸相同时，铝的电阻约为铜的一倍半。

四、倍数的减少

1) reduce by N times
   reduce N times
   reduce to N times
   reduce N times as much (many…) as
   N-fold reduction
   N times less than
   结构的译法。

上述所有结构均表达 "减少了 $\frac{n-1}{n}$" 或 "减少到 $\frac{1}{n}$"，所列结构中，除 reduce 外，尚有若干表示 "减少" 的同义词和近义词，如 decrease, shorten, drop, step down, cut down 等。在 N times less than 结构中，less 可换成其他弱比较级的词，如 lighter, weaker, shorter 等。

The new equipment *will reduce* the error probability *by seven* times.
新设备的误差概率将降低 6/7。

The voltage *has dropped five times.*
电压降低了 4/5。

2）half as much as
    twice less than   } 结构的译法。

上述 2 种结构均可译为 "比……少一半" 或 "比……少 1/2"。

The power output of the machine is *twice less than its* input.
该机器的输出功率比输入功率小 1/2。

五、分数和百分数增减的译法

1）分数和百分数的增减，一般表示净增减的部分，不包括底数在内。因此可直译成 "增加（减少）几分之几"，"增加（减少）百分之几"。

The pressure *will be reduced to one-fourth* of its original value.
压力将减少到原来数值的四分之一。

New booster can *increase* the pay load by 120%.
新型助推器能使有效负荷增加 120%。

2）X% + ""n
    v. + to + X%   } 结构的译法。

上述 2 种结构表示增减后的结果，包括底数在内。可直接译出或按倍数的译法处理。

This year the factory has produced 250% *the number of color TV sets* in 1985.
该厂今年生产的彩电是 1985 年的 250%。或者：该厂今年生产的彩电是 1985 年的 2.5 倍。或者：该厂今年生产的彩电比 1985 年增加了一倍半。

Reading Material

The Motivational Environment

Everything that surrounds us is part of our motivational environment. Depending upon our environment, we are motivated differently at a given point in time. We actually exist in what we call "micro-motivational" environments. These micro-environments make up the sum total of our motivational environment and are comprised of our work environment, family environment, social environment, team environment, peer environment, or even a nonfunctional environment.

Any one or all of these micro-environments can have an impact on the other. The negative impact of one of a person's micro-environments may cause that person to also react negatively in another one of his or her environments; this can happen even when the environment is, in itself, a positive one. For example, if an individual has problems at home, it may, and many times does, cause that highly motivated employee to become less safety conscious or productive at work.

In illustrating the complexity of this issue, let's think for a moment about employees problem. Many times these employees ask to be moved to a different job because they are either dissatisfied or are performing poorly in their current job. Amazingly, once the reassignment is made, their performance vastly improves. It is almost as if this worker becomes a different individual. When they are put into a new and different environment, they get a "new spark" and the new environment becomes their positive motivator; they've been revitalized! Many individuals do not like change but all of us react and are energized both negatively and positively by change. So you will need to make changes in your motivational approach when you see motivation waning.

1. Structuring the Motivational Environment

It takes some degree of organization and commitment to structure an environment where workers will be motivated to perform their work in a safe and healthy manner. The safety and health environment must have a foundation. A written safety and health program is the key component in providing and structuring that foundation. This written program should set the tone for safety and health within the work environment.

Firstly, you must explain and clarify the safety and health performance expectations. You cannot assume that workers know what is expected of them unless you tell them. You must make sure that your expectations are concise and consistent. If you want your workplace to be the safest in your industry then you must not deviate from what you expect. Tell people how you hope to achieve your expectation and do not fail to ask for advice on how to attain this expectation from everyone in your workforce. Always remember that people have a definite need to know. They certainly like to know what is going on.

Secondly, a way to keep your expectation out front regarding safety and health is to establish attainable and reachable safety and health goals. Goals which are understandable by all are much better motivators than ones which workers do not understand. It is good to involve those who will be impacted by the goal in the development of that goal. A goal to reduce our injury rate by 20 percent may sound fine to you but many workers do not know what an injury rate is or the components which go into calculating it. A better goal would be to keep the number of injuries below 10 per month. This can be easily tracked and counted on a monthly basis. The progress towards this goal can be posted regularly. All of us are goal driven, whether we recognize it or not. I venture to say that most of what an organization, team, individual accomplishes is the result of a goal set.

A third key to motivation is providing feedback. If you post the number of injuries each month, workers are being given feedback on the progress towards that goal. We need to know how we are doing in order to maintain our focus and motivation toward an outcome. Providing feedback is vitally important. How many times have you heard someone say, "I wish they would tell me how I am doing. I don't care if it is bad or good. I just want to know."

That person is saying, "Talk to me, please give me feedback."

Reinforcement is an important key in the motivational process. How you reinforce safety performance will determine whether it is strengthened or weakened. Reinforcers can be verbal feedback, a reward, or a consequence. It depends upon how the reinforcement is being used to drive home the message of accomplishment or failure to reach the safety and health goal. Telling a worker that you really appreciate the safe way he is performing his job is feedback which reinforces the type of behavior which you desire and fosters motivation in that individual. Reinforcement for safety and health needs to be more frequent than once a year. Monthly would be best but quarterly is also adequate. Unless the reward is very large a year is too long a period to have to wait for reinforcement.

A key which has been discussed earlier is involvement. We should make every effort to involve workers who have a vested interest and invest their energy towards the outcome of a safe and healthy workplace. There are many ways to involve workers in the safety and health initiative. These range from participation in a safety and health committee to conducting inspections. This involvement needs to be nurtured and recognized. When our contributions are supported and recognized it has a positive impact upon our behavior and thus we are more motivated. You may need to be very creative in finding ways to get workers involved.

Many times workers like to be able to monitor their own progress towards a goal or expectation. If you have a chalkboard, workers themselves can mark the board each time that they or a fellow worker are injured on the job. Even if somewhat inaccurate, workers need to sense that they are involved in the process of safety and health at their workplace. This can lead to more team work and more motivation in the workplace. Self-monitoring may not always be an option but do not overlook its impact when it can be used.

One of the most debated keys to motivation in the safety arena is rewards. In all aspects of life we tend to focus and perform better when rewards are involved. Rewards are not a quick fix to problems with your company's safety and health performance. Rewards are a complement to the safety and health initiative. If you do not have all the nuts and bolts of a safety and health program in place then rewards are not a replacement for failure to effectively manage the safety and health effort. I would suggest to you that rewards are the icing on the cake. If your company is not performing well to your safety and health goals then rewards may be used to keep workers reinforce performance, or recognize the attainment of goals and expectations. Employers often say, "I already pay my workforce to work safely," and I say to them, "Then why in the world do you give them production bonuses when you already pay them to produce?" Of course, the answer is that it gives them a little more motivation. I do hate to think that a person's safety can be bought and it would not be in their best interest to become injured or ill. But, on the other hand, it is my sense that using rewards as motivators, reinforcers, and reminders for a higher purpose than that of trying to provide the best approaches to obtain a safe and healthy workplace is appropriate.

One caution with rewards which I would make you aware of is that money is a one-time occurrence and the reason that one receives it is quickly forgotten when it is spent. Also, any reward must be of value to the person receiving it. A gift certificate to Fashion Bug may not be viewed as valuable to your male employees. This is why if gifts of some sort are to be given out you might want a variety available through a catalog which would appeal to a wider range of your workforce. You could provide catalog dollars which could increase in numbers as progress toward your goals is attained.

Rewards need to be tangible so that when the person sees it in his home or workplace it serves as a reminder of why he received it. Appropriate rewards might include a bond, a plaque, a pin with a one on it (one year without an accident), a certificate, an embossed hat, jersey, jacket or other assorted items which are reminders to the individual to stay motivated towards the safety and health performance goals.

The organized approach to safety and health should address each of the previous eight keys. Within these paragraphs you find these keys to motivation which should be an integral part of the occupational safety and health prevention program initiative.

Many tangible and intangible factors comprise the motivational work environment. Something that is tangible could be something as simple as moving a piece of equipment in order to create a more desirable environment or granting a request. Something intangible could be your ability to change the way someone feels about you.

When it comes to developing a safe and productive motivational work atmosphere, the intangible motivational issues are just as important as the tangible ones, but they are also the most time consuming. These challenges run a broad spectrum and, just to list a few, could be some of the following examples: Changing the way a person is treated by peers, colleagues, or supervisors; helping an individual gain a positive perception of the environment; or developing a new and positive attitude towards workplace safety and health.

You should develop an environment where the majority will be positively motivated to perform but this should not prohibit you from making adjustments, when possible, to address individual needs. Furthermore, be sure that you don't allow too much flexibility (i.e., favoritism, etc.) or it could destroy a good situation for the majority.

To assure equal treatment, require all to abide by the rules. For instance, when there is a set group of safety and health rules for your workplace, you should never allow one person to abuse these rules while holding others rigidly to them; this will cause disenchantment with safety and health issues. Top management, supervisors, foremen, and workers should be treated equally and fairly by requiring them to comply with the safety and health rules and policies. As an example of this, while I was working with one company the safety director wore moccasins while everyone else was required to wear hard-toed shoes. This type of behavior should not be acceptable for one person and not others. Although you may be the head of an organization, or the "boss", you should never consider yourself so lofty that you do not adhere to your own safety and health rules and policies. It is so very important that management and supervisors set the tone of the work environment in regards to safety and health on the job. When setting up an environment where you want those involved to be motivated, you should first address the physical needs. For example, in the work environment there may be the need to provide the proper tools and personal protective equipment in order for the workers to do their work safely.

Your ability to structure an environment which provides individual needs and adequate stimulus to motivate each person to his or her "full capacity" is desirable but not usually possible. In fact, you actually have little chance of setting up the "perfect" environment for every person. There are just too many other environments and factors, which compete with you and what you desire each individual to accomplish. However, do the best you can for each person and then each individual will make a conscious decision as to whether he or she wants to perform safely in the workplace. This is the reason that each worker should know the conseuences of any unsafe performance. You should develop mechanisms to assist these individuals to perform safely, but also have disciplinary procedures for those who elect not to comply with the safety and health rules. As part of setting the environment, be assured that each worker understands the expectations regarding working safely. It is also useful to involve them in setting the safety and health rules and goals and to know the expected outcomes. Each worker needs to understand that there will be consequences or penalties for disregarding or violating the safety and health requirements of their work. Therefore, goals are important in setting performance objectives for the company's safety and health program. Track the progress of the safety and health goals and provide feedback. This allows the workers to monitor their own accomplishments in their work area. Recognize the workers who are progressing towards the safety and health goals and reinforce safe work behaviors.

2. Reacting to the Motivational Environment

You can provide all of the bells and whistles but, if you do not pay attention to some fundamental characteristics of people, you will not be successful in developing a good motivational environment. Some of the fundamental principles you need to be aware of when working with people are:

- Individuals view themselves as very special. Thus, praise, respect, responsibility, delegated authority, promotions, recognitions, bonuses, and raises add to their feelings of high self-esteem and need to be considered when structuring a motivational environment.
- Instead of criticism, use positive approaches and ask for corrected behavior. Individuals usually react in a positive manner when this approach is used.
- Verbally attacking (disciplining) individuals tends to illicit a very defensive response. Therefore, it is better to give praise in public and, when necessary, criticize in private.
- Individuals are unique and given the proper environment, which will astound you with their accomplishments and creativity (even those individuals whom you consider non-creative).

Remember, the final outcome lies with the employees; they will decide whether or not to perform safely. But, if the employer has done his or her part, workers will not be able to hold you responsible for the decisions they have chosen to make.

There will be people who elect to work unsafely even though the environment may be very motivational to the majority. Thus, when discussing work, you will need to pay close attention to the motivational environment, and work at making it the very best! But when there are those who do fail to perform safely, there should be consequences and discipline administered quickly and fairly. If there is no one enforcing the speed limit, then who will abide by it? Either enforce the rules or lose the effectiveness of your motivational effort. The key to a successful motivational environment is to pique the interest of people. Let them know you want them to succeed; give them responsibility; and leave them alone to accomplish those goals and succeed. If the above principles are not taken into consideration when setting up your motivational environment, you will be more likely to encounter problems with your success rate. As an illustration, a supervisor noticed that his workers were not giving him the performance he expected. He was having difficulty receiving top quality written reports from them and, therefore, had been rewriting each report. When the supervisor was asked if his employees were aware that he was rewriting their reports and, if so, did he think they were putting forth their best effort, he answered, "Oh yes." But after thinking this question through he decided to go back to his work area and ask his workers the same question that was asked to him. Their reply to him was, as expected, that they were only giving a half-hearted effort since they knew the report would be rewritten. As you can see from this example, you need to be cautious that you don't set yourself up for this type of response.

The way that you structure the motivational environment will allow individuals within the work groups to accomplish safety and health goals and assure that they are free from injury and illness. What you need to do is set up an environment where people can be successful. And, in order for that environment and the people within it to succeed, you must demonstrate that you genuinely care about them and the purpose of the mission (goal) they are trying to attain, which in this case is a safe and healthy workplace. Next, you need to be open to learning from your own experiences, as well as from others. This will facilitate flexibility in your encounters and give you the ability to make the necessary changes. You need to be honestly perceived by everyone as working diligently to prevent workplace incidents and be willing to work at motivating those who are not in tune with your safety and health attempts. It does take an added effort to motivate others.

It is imperative that you realize when you have reached a point where you have accomplished as much as you can and have lost the effectiveness of the safety and health environment that you have structured. This may be an indicator that you need to change your approach. As an example of this type of situation, think for a moment about coaches, especially those who are in the professional ranks or at larger collegiate institutions. When they become ineffective, they are forced to move onto other coaching positions. But once they are in their new position, and even though they had become ineffective in their previous one, they often are able to rejuvenate a program, which, until their arrival, was unsuccessful. In these cases we realize that the coaches are still the same people but they become ineffective because the environment changed in their previous jobs and they were unable to adapt to those changes. Nevertheless, when they were introduced into a new position, they once again became successful.

Thus, when structuring your motivation environment, be sure to load it with as many of these true motivators as possible. They are the most successful incentives and encourage consistent and improved safe performance. Some motivational environment examples are as follows:

In the past, companies have tried to motivate people by reducing the hours worked, giving longer vacations, increasing wages, increasing benefit packages, providing career counseling services, training supervisors in communications, and organizing interactive groups. However, these incentives have not proven to be highly effective in increasing productivity. Therefore, it is important that we determine what affects the satisfaction or dissatisfaction on a job or, for that matter, anywhere else.

In structuring a motivational environment, it is important to help people grow and learn through the task they are asked to perform. Prepare them to stretch their abilities to new and more difficult tasks and help them advance to higher levels of achievement. Help them use and recognize their unique abilities and make sure they can see the results of their efforts. Be sure and recognize when a task is well done; give a promotion or award and provide or reinforce performance with constructive feedback. This is not only applicable at the workplace but is also standard for life situations whether it be school, sports, home, social groups or peer groups.

In recent years companies such as Ford, Volvo, and General Motors, as well as many others, have found that team approaches to the work environment are very effective. They have found that an increase in quality and overall job satisfaction transpires when a work group is assigned a task and then given control over such decisions as who does what tasks, how the tasks will be accomplished, and who has the authority to stop the process if quality is in question.

With this type of system in force, the supervisor is no longer responsible for completion of the task; the group has that responsibility and control. The supervisor's main duty becomes one of advising, providing feedback, and assuring that all materials and tools are available to accomplish the job. This approach has also been very successful with quality circles but may not work in all environments since the end product is not the same for all individuals and in all situations.

When there are barriers which keep you from being able to set up a good motivational setting, put an even greater emphasis on the non-tangibles (recognition, achievement, responsibility and challenge). As an example, let us consider the M. A. S. H. television series. The physical setting was terrible, the wounded were disheartening, and the tools needed to accomplish their mission were often missing. But, discipline was not stringent, protocol was lax, individuality and recognition were endeared and this made the mission not only challenging but also rewarding.

As you can see from these examples, you cannot always predict the way in which individuals will react to a motivational environment, but you can predict with some certainty that if there is no attempt to set up a good motivational environment, an integral part of motivation will be lost. Thus, with this piece of the puzzle missing, the other facets of the motivational plan cannot be effectively applied.

New Words and Expressions

consistent [kən'sɪstənt]   adj. 一致的，调和的
vested interest   n. 特权阶级；既得利益
bonus ['bəʊnəs]   n. 奖金，红利
tangible ['tændʒəbl]   adj. 切实的
plaque [plɑːk]   n. 在墙上作装饰或纪念用的薄金属板或瓷片
spectrum ['spektrəm]   n. 系列，范围，幅度
abide [ə'baɪd]   v. 坚持，遵守；忍受，容忍
disenchantment [ˌdɪsɪn'tʃɑːntmənt] n. 觉醒，清醒
moccasin ['mɒkəsɪn]   n. 鹿皮鞋；软拖鞋
lofty ['lɔ(ː)fti]   adj. 高高的；崇高的；高傲的；高级的
adhere to   坚持；追随，拥护
reinforce [ˌriːɪn'fɔːs]   v. 加强，增援，补充
illicit [ɪ'lɪsɪt]   adj. 违法的
astound [əs'taʊnd]   v. 使惊骇，使大吃一惊
pique [piːk]   v. 伤害……的自尊心，使……生气
rejuvenate [rɪ'dʒuːvɪneɪt] v. 使复原，使更新，使人恢复精神，使恢复活力
transpire [træns'paɪə]   v. 发生，得知，发现

Unit Nine

Accident Investigations

Although accident investigation is an after-the-fact approach to hazard identification, it is still an important part of this process. At times hazards exist, which no one seems to recognize until they result in an accident or incident. In complicated accidents it may take an investigation to actually determine what the cause of the accident was. This is especially true in cases where death results and few or no witnesses exist. An accident investigation is a fact-finding process and not a fault-finding process with the purpose of affixing blame. The end of any result of an accident investigation should be to assure that the type of hazard or accident does not exist or occur in the future.

Your company should have a formalized accident investigation procedure, which is followed by everyone. It should be spelled out in writing and end with a written report using as a foundation of your standard company accident investigation form. It may be your workers' compensation form or an equivalent from your insurance carrier.

Accidents and even near misses should be investigated by your company if you are intent on identifying and preventing hazards in your workplace. Thousands of accidents occur throughout the United States every day. The failure of people, equipment, supplies, or surroundings to behave or react as expected causes most of the accidents. Accident investigations determine how and why these failures occur. By using the information gained through an investigation, a similar or perhaps more disastrous accident may be prevented. Accident investigations should be conducted with accident prevention in mind. Investigations are NOT to place blame.

An accident is any unplanned event that results in personal injury or in property damage. When the personal injury requires little or no treatment, it is minor. If it results in a fatality or in a permanent total, permanent partial, or temporary total (lost-time) disability, it is serious. Similarly, property damage may be minor or serious. Investigate all accidents regardless of the extent of injury or damage. Accidents are part of a broad group of events that adversely affect the completion of a task. These events are incidents. For simplicity, the procedures discussed in later sections refer only to accidents. They are, however, also applicable to incidents.

1. Accident Prevention

Accidents are usually complex. An accident may have 10 or more events that can be causes. A detailed analysis of an accident will normally reveal three cause levels: basic, indirect, and direct. At the lowest level, an accident results only when a person or object receives an amount of energy or hazardous material that cannot be absorbed safely. This energy or hazardous material is the DIRECT CAUSE of the accident. The direct cause is usually the result of one or more unsafe acts or unsafe conditions, or both. Unsafe acts and conditions are the indirect causes or symptoms. In turn, indirect causes are usually traceable to poor management policies and decisions, or to personal or environmental factors. These are the basic causes.

In spite of their complexity, most accidents are preventable by eliminating one or more causes. Accident investigations determine not only what happened, but also how and why. The information gained from these investigations can prevent recurrence of similar or perhaps more disastrous accidents. Accident investigators are interested in each event as well as in the sequence of events that led to an accident. The accident type is also important to the investigator. The recurrence of accidents of a particular type or those with common causes shows areas needing special accident prevention emphasis.

2. Investigative Procedures

The actual procedures used in a particular investigation depend on the nature and results of the accident. The agency having jurisdiction over the location determines the administrative procedures. In general, responsible officials will appoint an individual to be in charge of the investigation. An accident investigator should use most of the following steps:

- Define the scope of the investigation.
- Select the investigators. Assign specific tasks to each (preferably in writing).
- Present a preliminary briefing to the investigating team.
- Visit and inspect the accident site to get updated information.
- Interview each victim and witness. Also interview those who were present before the accident and those who arrived at the site shortly after the accident. Keep accurate records of each interview. Use a tape recorder if desired and if approved.
- Determine the following:
  - What was not normal before the accident.
  - Where the abnormality occurred.
  - When it was first noted.
  - How it occurred.
- Determine the following:
  - Why the accident occurred.
  - A likely sequence of events and probable causes (direct, indirect, basic).
  - Alternative sequences.
- Determine the most likely sequence of events and the most probable causes.
- Conduct a post-investigation briefing.
- Prepare a summary report including the recommended actions to prevent a recurrence. Distribute the report according to applicable instructions.

An investigation is not complete until all data are analyzed and a final report is completed. In practice, the investigative work, data analysis, and report preparation proceed simultaneously over much of the time spent on the investigation.

3. Fact-Finding

Investigator collects evidence from many sources during an investigation, gets information from witnesses and observation as well as by reports, interviews witnesses as soon as possible after an accident, inspects the accident site before any changes occur, takes photographs and makes sketches of the accident scene, records all pertinent data on maps, and gets copies of all reports. Documents containing normal operating procedures flow diagrams, maintenance charts or reports of difficulties or abnormalities are particularly useful. Keep complete and accurate notes in a bound notebook. Record pre-accident conditions, the accident sequence and post-accident conditions. In addition, document the location of victims, witnesses, machinery, energy sources, and hazardous materials.

In some investigations, a particular physical or chemical law, principle, or property may explain a sequence of events. Include laws in the notes taken during the investigation or in the later analysis of data. In addition, gather data during the investigation that may lend itself to analysis by these laws, principles, or properties. An appendix in the final report can include an extended discussion.

4. Interviews

In general, experienced personnel should conduct interviews. If possible, the team assigned to this task should include an individual with a legal background. After interviewing all witnesses, the team should analyze each witness' statement. They may wish to re-interview one or more witnesses to confirm or clarify key points. While there may be inconsistencies in witnesses' statements, investigators should assemble the available testimony into a logical order. Analyze this information along with data from the accident site.

Not all people react in the same manner to a particular stimulus. For example, a witness within close proximity to the accident may have an entirely different story from one who saw it at a distance. Some witnesses may also change their stories after they have discussed it with others. The reason for the change may be additional clues.

A witness who has had a traumatic experience may not be able to recall the details of the accident. A witness who has a vested interest in the results of the investigation may offer biased testimony. Finally, eyesight, hearing, reaction time, and the general condition of each witness may affect his or her powers of observation. A witness may omit entire sequences because of a failure to observe them or because their importance was not realized.

5. Report of Investigation

As noted earlier, an accident investigation is not complete until a report is prepared and submitted to proper authorities. Special report forms are available in many cases. Other instances may require a more extended report. Such reports are often very elaborate and may include a cover page, title page, abstract, table of contents, commentary or narrative discussion of probable causes, and a section on conclusions and recommendations.

Unit Nine Accident Investigations

Accident investigation should be an integral part of your written safety and health program. It should be a formal procedure. A successful accident investigation determines not only what happened, but also finds how and why the accident occurred. Investigations are an effort to prevent a similar or perhaps more disastrous sequence of events. You can then use the resulting information and recommendations to prevent future accidents.

Keeping records is also very important to recognizing and reducing hazards. A review of accident and injury records over a period of time can help pinpoint the cause of some accidents. If a certain worker shows up several times on the record as being injured, it may indicate that the person is physically unsuited for the job, is not properly trained, or needs better supervision. If one or two occupations experience a high percentage of the accidents in a workplace, they should be carefully analyzed and countermeasures should be taken to eliminate the cause. If there are multiple accidents involving one machine or process, it is possible that work procedures must be changed or that maintenance is needed. Records that show many accidents during a short period of time would suggest an environmental problem.

Once the hazards have been identified then the information and sources must be analyzed to determine their origin and the potential to remove or mitigate their effects upon the workplace. Analysis of hazards forces us to take a serious look at them.

New Words and Expressions

affix [əˈfɪks]	v.	归咎于，归于
spell out		使十分清楚明白
compensation [ˌkɔmpenˈseɪʃən]	n.	补偿，赔偿
disastrous [dɪˈzɑːstrəs]	adj.	损失惨重的
fatality [fəˈtælɪti]	n.	不幸，灾祸
applicable [ˈæplɪkəbl]	adj.	可适用的，可应用的
symptom [ˈsɪmptəm]	n.	症状，征兆
traceable [ˈtreɪsəbl]	adj.	可追踪的，起源于……的
recurrence [rɪˈkʌrəns]	n.	复发，重现，循环
jurisdiction [ˌdʒʊərɪsˈdɪkʃən]	n.	权限
scope [skəʊp]	n.	（活动）范围
preferably [ˈprefərəbli]	adv.	更适宜
briefing [ˈbriːfɪŋ]	n.	简报
approved [əˈpruːvd]	adj.	经核准的，被认可的
abnormality [ˌæbnɔːˈmælɪti]	n.	变态，畸形，异常性
simultaneously [ˌsɪməlˈteɪnɪəsly]	adv.	同时地
sketch [sketʃ]	n.	略图，草图
pertinent [ˈpɜːtɪnənt]	adj.	有关的，相关的；中肯的
document [ˈdɒkjʊmənt]	v.	证明
inconsistency [ˌɪnkənˈsɪstənsi]	n.	矛盾
testimony [ˈtestɪməni]	n.	陈述；证词
proximity [prɒkˈsɪmɪti]	n.	接近，临近
traumatic [trɔːˈmætɪk]	adj.	外伤的，创伤的

vested ['vestɪd]　　　　　　　　adj. 既定的
biased ['baɪəst]　　　　　　　　adj.（统计试验中）结果偏倚的
pinpoint ['pɪnˌpɔɪnt]　　　　　　v. 查明
mitigate ['mɪtɪgeɪt]　　　　　　v. 减轻

Translation Skill

科技英语翻译技巧（八）——被动语态的译法

英语和汉语都有被动语态，但两种文字对被动语态的运用却不尽相同。语态的使用不同，语气和情调也随之而异。在英译汉时，必须注意被动语态的译出，不要过分拘泥于原文的被动结构。

在英语科技书刊中，被动语态句子十分常见，主要有以下几种情况：

1）当不必说出主动者或无法说出主动者时。

For a long time aluminum *has been thought* as an effective material for preventing metal corrosion.

长期以来，铝被当作一种有效地防止金属腐蚀的材料。

2）为了强调被动者，使其位置鲜明、突出时。

Three machines *can be controlled* by a single operator.

三台机器能由一个操作者操纵。

3）为了更好地联接上下文。

Vulcanized rubber was a perfect insulating material; on the railway it *was used* for shock-absorbers and cushions.

硫化橡胶是一种很好的绝缘材料，在铁路上可用来做减振器和减振垫。

科技英语主要是叙述事理，往往不需要说出主动者，或对被动者比主动者更为关心。此外，科技工作者为了表示客观和谦虚的态度，往往避免使用第一人称，因而尽可能使用被动语态。因此，在翻译英语被动语态时，大量语句应译成主动句，少数句子仍可译成被动句。

一、译成汉语的主动句

1. 原主语仍译为主语

当英语被动句中的主语为无生命的名词，又不出现由介词 by 引导的行为主体时，往往可译成汉语的主动句，原句的主语在译文中仍为主语。这种把被动语态直接译成主动语态的句子，实际是省略了"被"字的被动句。

If a machine part is not well *protected*, it will become rusty after a period of time.

如果机器部件不好好防护，过一段时间后就会生锈。

Every moment of every day, energy *is being transformed* from one form into another.

每时每刻，能量都在由一种形式转换成另一种形式。

2. 把原主语译成宾语，而把行为主体或相当于行为主体的介词宾语译成主语

Friction can be reduced and the life of the machine prolonged by *lubrication*.

润滑能减少摩擦，延长机器的寿命。

Modern scientific discoveries lead to the conclusion that energy may be created from matter and that *matter* in turn, may be created from energy.

近代科学的发现得出这样的结论：物质可以产生能量，能量又可以产生物质。

3. 在翻译某些被动语态时，增译适当的主语使译文通顺流畅

1）增译逻辑主语：原句未包含动作的发出者，译成主动句时可以从逻辑出发，适当增加不确定的主语，如"人们""有人""大家""我们"等，并把原句的主语译成宾语。

To explore the moon's surface, rockets *were launched* again and again.

为了探测月球的表面，人们一次又一次地发射火箭。

Although the first synthetic materials were created little more than a hundred years ago, they *can be found almost* everywhere.

虽然第一批合成材料仅在100多年前才研制出来，但现在人们几乎到处都能见到它们。

2）某些要求复合宾语的动词，如 believe, consider, find, know, see, think 等，用于被动语态时，翻译时往往可加上不确定主语。

Salt is *known* to have a very strong corroding effect on metals.

大家知道，盐对金属有很强的腐蚀作用。

3）由 it 做形式主语的被动句型。这种句型在科技英语中比比皆是，十分普遍，汉译时一般均按主动结构译出。即将原文中的主语从句译在宾语的位置上，而把 it 做形式主语的主句译成一个独立语或分句。

It is believed to be natural that more and more engineers have come to prefer synthetic material to natural material.

越来越多的工程人员宁愿用合成材料而不用天然材料，人们相信这是很自然的。

二、译成汉语的其他句型

1. 译成汉语的无主句

英语中许多被动句不需或无法讲出动作的发出者，往往可译成汉语的无主句，而把原句中的主语译成宾语。英语中有些固定的动词短语，如：make use of, pay attention to, take care of, put an end to 等用于被动句时，常译成被动句。

In the watch making industry, the tradition of high precision engineering *must be kept*.

在钟表制造业中，必须保持高精度工艺的传统。

Attention *has been paid to* the new measures to prevent corrosion.

已经注意到采取防腐新措施。

2. 译成汉语的判断句

凡着重描述事物的过程、性质和状态的英语被动句，实际上与系表结构很相近，往往可译成"是……的"结构。

The voltage *is not controlled* in that way.

电压不是用那样的方法控制的。

3. 译成汉语的被动句

英语中有些着重被动动作的被动句，要译成被动句，以突出其被动意义。被动含义可用"被""由""给""加以""为……所""使""把""让""叫""为""挨""遭"等表达。

The metric system *is* now *used* by almost all countries in the world.

米制现在被全世界几乎所有的国家采用。

Reading Material

Cooperation between Insurance and Prevention

Work injuries are an unwelcome byproduct of economic activity. In part, they are random events, but they are all, to some extent, under the control of workers and employers. Employers can reduce the number of workplace injuries and illnesses by investing in safer technologies, providing workers with personal safety protective equipment, training workers and their supervisors; workers can avoid accidents by following safe work practices and by taking greater care on the job.

Both parties incur costs when an accident occurs. Workers' costs include potential loss of income and medical expenses associated with treatment and rehabilitation as well as intangibles, such as pain and suffering and disability that reduce the ability to enjoy life. Employers' costs include interruptions in production and damage to capital equipment and physical plant.

Since accident prevention also entails costs to employers and employees, public policy should encourage employers and employees to optimize the allocation of the combined costs of accidents and accident prevention that are incurred by both workers and employers. It is possible to spend either too much or too little on accident prevention. Investment in accident prevention is socially efficient when total costs are minimized, that is when one additional dollar spent on prevention reduces accident costs by exactly one dollar. When the majority of the economic costs of the preventable occupational health and safety (OHS) burden are borne by parties external to the firm, firms will not have a clear understanding of actual costs of under-investment in workplace health and safety. Insurance instruments can clarify the precise economic costs incurred by firms. Regulation and enforcement can raise the firm-level costs of under-investment in occupational health and safety. And information and consultation services provided by prevention authorities can increase firm-level recognition of effective OHS policies and practices.

There are at least three ways to the rule of OHS. The first way, classically identified as OHS regulation, involves the promulgation of rules prescribing or proscribing specific policies and practices by employers, which are enforced through onsite inspections and monetary penalties for infractions. Direct rule attempts to change employer behavior by prescribing specific practices. Regulatory sanctions have the effect of raising the level of "efficient" health and safety investment by the firm to the expected value of the sanction. In so doing, regulation introduces costs. There are two costs that must be considered by the efficient regulator: the administration costs of regulation and the cost of regulatory error. The evidence for the effectiveness of regulatory inspection and enforcement in occupational.

The second way emphasizes the role of economic incentive that reward or punish employers on the basis of safety and health outcomes rather than behaviors. This way is embodied in experience rating. At the company level, experience-rated workers' compensation insurance premiums present a choice at the company level between investments in accident prevention and investments in disability management. Higher spending on diability management will generally mean lower spending on accident prevention.

Third way, termed internal responsibility, is designed to improve safety and health conditions through workers empowerment and involves three principal elements: ① workers' right to refuse unsafe work,

②workers' rights to information on nature of workplace hazards and ③joint labour-management safety and health committees. The emphasis on the free exchange of high quality information gives expression to a key provision in economic theory; efficient decision-making within firms requires an understanding of the economic costs and benefits of different courses of action, including the costs of doing nothing compared to the costs of effective methods of removing a risk or hazard.

The premise behind the economic theory of experience-rated insurance is that it provides more precise information to individual firms concerning the economic cost of occupational injury and illness than ways to levying insurance premiums based on collective experience of all employers in an economic sector or a national labour market. Unfortunately, it is very frequently the case that insurance premiums schedule only a fraction of the true direct and indirect cost of workplace injury and illness. The majority of the economic costs of occupational injury and illness are externalized. As a result of the broad failure to "internalize" the full costs of occupational injury and illness at the firm-level, the conventional guidance of economic theory, that the firm will invest in health and safety up to a threshold less than or equal to the firms' costs arising from occupational injury and illness can be expected to yield a response which is not sufficient to mitigate the true costs of occupational injury and illness.

In theory, providers of insurance have some incentives to improve loss management practices among holder of insurance policies. Improved loss management practices among insurance policy holders should reduce the range of error in risk assessment on the part of insurance providers. There are many examples of insurance providers delivering loss management services: automobile driver education, home or commercial property's fire protection practice, workplace health and safety practice.

While some have argued that the profit motive provides private insurance carriers with an incentive to control losses and provide insured parties with safety management service, others have suggested that loss management services are provided by insurers to create a premium service brand identity in competitive markets. Whatever the actual case, there is broad consensus that private insurance carriers will only provide loss management services up to the point where an additional dollar spent on loss control is equal to the additional dollar of losses saved.

Public insurance providers, such as the single-payer monopoly workers' compesation agencies, may be more likely to embrace a dual mandate; both to operate an efficient and fiscally sound insurance program and to expend resources or invest capital to reduce the future burden of preventable morbidity in workplace. The latter objective is a social objective, and is external to the function of insurance.

Insurance providers in private markets are not in the business of reducing or eliminating property and personal economic losses due to accidents or injuries. Private market insurance is specifically in the business of selling insurance policies at prices which cover the economic risks borne by the insurance policies. Private market insurance is, then, indifferent to the scale of societal economic loss arising from accidents and injuries.

Given the high proportion of true costs of occupational injury and disease that are borne by parties external to the firm, there is a stronger theoretical justification for the intervention of prevention authorities to influence the decision-making of employers. As reviewed in this presentation, there are three types of instruments available to prevention authorities to influence firms: regulation, inspection and enforcement, insurance incentives and information and consultation (including the sponsorship of research investments).

In a great many jurisdictions, especially those jurisdictions with publicly mandated single-payer insurance authorities, we have seen substantial experimentation by the insurance function as a direct funder of prevention activities. Much of this experimentation has been at a modest scale of financial commitment.

A promising area for future program innovation may be for insurance funds to act as providers of capital for firm-level or sector level investments in technology which improves the future health of workers. Capital is scarce, and there is intense competition within enterprises for access to these scarce resources. Insurers have large capital reserves, of which a substantial share is invested (risked) in investment markers. Investment managers may well benefit from comparing capital market returns to the returns arising from programs which support capital investments to reduce the future economic burden of occupational injury and disease.

Peter Dorman, a thoughtful commentator on the economic principles underlying OHS, has offered this additional observation: expenditures by firms on improvements in working conditions are investments in strict economic sense—they are costs borne in an earlier period in order to reap benefits in later ones. But investments must be financed. For large enterprises this may not be a problem, since they may have sufficient internal finance to meet all reasonable OHS needs. Smaller firms, however, must often turn to external source of funding. The cost and availability of finance is crucially dependent on the degree of collateralization—the ability of borrowers to put up assets as security behind their pledge to repay. Typically, the loans of which investments are made are collateralized by the assets the investments purchase or produce, such as equipment, materials, patents and stocks of finished or semi-finished goods. This does not work, in every case, for OHS investments, however, because the asset may well be the workforce itself, and workers cannot be offered as collateral. All investments in human capital, including investments in OHS, are subject to adverse reverse discrimination in financial markets.

Policies that move in the direction of more aggressive prevention expenditures by insurance agencies will encounter resistance. Some of this resistance arises simply from the conservative orientation of most organizations to "stick to our knitting". But there are also longstanding conceptual premises underlying the purposes of insurance which argue that prevention expenditures are not the business of the insurance function. This perspective would ask: Why not simply set the incentives for employers to invest in health and safety through the mechanism of the insurance premium, and leave the decision-making to the firm? Some responses might include:

- that the optimal solution may be at sector, rather than firm level;
- that insurers may have better information than individual employers as to the most efficient balance of investments/economies of scale/expertise in knowledge;
- insurers may be able to invest a larger amount than individual firms as a function of a longer horizon to capture the return on the investment;
- and insurers may be more capable as innovators.

As the efforts of firms, insurers and prevention authorities continue to make progress in reducing the preventable burden of occupational injury and illness, economic losses as a share of national GDP will decline. We might predict, however, that the relative share of resources devoted to compensating the economic losses of workers and the share devoted to ensuring effective prevention will need to shift. This will occur naturally even in a most conservative scenario in which the total economic value of prevention services remains constant while the economic value of compensation for economic losses declines. But it may be useful to note that the resources required to prevent the first occupational injury or disease will be much less than the resources required to prevent the last injury of occupational illness. As we move into the future, prevention authorities will need to claim increasing resources to achieve the social objective of eliminating the preventable burden of occupational injury and disease.

New Words and Expressions

rehabilitation	['riː(h)ə,bɪlɪ'teɪʃən]	n.	复原
intangible	[ɪn'tændʒəbl]	adj.	难以明了的，无形的
promulgation	[,prɒməlgeɪʃən]	n.	颁布
prescribe	[prɪs'kraɪb]	v.	指示，规定
proscribe	[prəʊ'skraɪb]	v.	禁止
infraction	[ɪn'frækʃən]	n.	违反，侵害
premium	['prɪmjəm]	n.	额外费用，保险费，奖金
monopoly	[mə'nɒpəli]	n.	垄断，专利权
jurisdiction	[,dʒʊərɪs'dɪkʃən]	n.	权限
pledge	[pledʒ]	n.	保证，誓言，抵押，抵押品

Unit Ten

Safety Electricity

Occupational Safety and Health Administration (OSHA) and state safety laws have helped to provide safe working areas for electricians. Individuals can work safely on electrical equipment with today's safeguards and recommended work practices. In addition, an understanding of the principles of electricity is gained. Ask supervisors when in doubt about a procedure. Report any unsafe conditions, equipment, or work practices as soon as possible.

1. Fuses

Before removing any fuse from a circuit, be sure the switch for the circuit is open or disconnected. When removing fuses, use an approved fuse puller and break contact on the hot side of the circuit first. When replacing fuses, install the fuse first into the load side of the fuse clip, then into the line side.

2. Groundfault Circuit Interrupts

A groundfault circuit interrupter (GFCI) is an electrical device which protects personnel by detecting potentially hazardous ground faults and quickly disconnecting power from the circuit. A potentially dangerous ground fault is any amount of current above the level that may deliver a dangerous shock. Any current over 8 mA is considered potentially dangerous depending on the path the current takes, the amount of time exposed to the shock, and the physical condition of the person receiving the shock.

Therefore, GFCIs are required in such places as dwellings, hotels, motels, construction sites, marinas, receptacles near swimming pools and hot tubs, underwater lighting, fountains, and other areas in which a person may experience a ground fault.

A GFCI compares the amount of current in the ungrounded (hot) conductor with the amount of current in the neutral conductor. If the current in the neutral conductor becomes less than the current in the hot conductor, a ground fault condition exists. The amount of current that is missing is returned to the source by some path other than the intended path (fault current). A fault current as low as 4 mA to 6 mA activates the

GFCI and interrupts the circuit. Once activated, the fault condition is cleared and the GFCI manually resets before power may be restored to the circuit.

GFCI protection may be installed at different locations within a circuit. Direct-wired GFCI receptacles provide a ground fault protection at the point of installation. GFCI receptacles may also be connected to provide GFCI protection at all other receptacles installed downstream on the same circuit. GFCI CBs (circuit breakers), when installed in a load center or panelboard, provide GFCI protection and conventional circuit overcurrent protection for all branch-circuit components connected to the CB.

Plug-in GFCIs provide ground fault protection for devices plugged into them. These plug-in devices are often used by personnel working with power tools in an area that does not include GFCI receptacles.

3. Electrical Shock

Strange as it may seem, most fatal electrical shocks happen to people who should know better. Here are some electromedical facts that should make you think twice before taking chances.

It's not the voltage but the current that kills. People have been killed by 100 volts AC in the home and with as little as 42 volts DC. The real measure of a shock's intensity lies in the amount of current (in milliamperes) forced through the body. Any electrical device used on a house wiring circuit can, under certain conditions, transmit a fatal amount of current.

Currents between 100 and 200 milliamperes (0.1 A and 0.2 A) are fatal. Anything in the neighborhood of 10 milliamperes (0.01 A) is capable of producing painful to severe shock.

As the current rises, the shock becomes more severe. Below 20 milliamperes, breathing becomes labored; it ceases completely even at values below 75 milliamperes. As the current approaches 100 milliamperes ventricular fibrillation occurs. This is an uncoordinated twitching of the walls of the heart's ventricles. Since you don't know how much current went through the body, it is necessary to perform artificial respiration to try to get the person breathing again; or if the heart is not beating, cardio-pulmonary resuscitation (CPR) is necessary.

Electrical shock occurs when a person comes in contact with two conductors of a circuit or when the body becomes part of the electrical circuit. In either case, a severe shock can cause the heart and lungs to stop functioning. Also, severe burns may occur where current enters and exits the body.

Prevention is the best medicine for electrical shock. Respect all voltages, have a knowledge of the principles of electricity, and follow safe work procedures. Do not take chances. All electricians should be encouraged to take a basic course in CPR (cardio-pulmonary resuscitation) so they can aid a coworker in emergency situations.

Always make sure portable electric tools are in safe operating condition. Make sure there is a third wire on the plug for grounding in case of shorts. The fault current should flow through the third wire to ground instead of through the operator's body to ground if electric power tools are grounded and if an insulation breakdown occurs.

4. First Aid for Electrical Shock

Shock is a common occupational hazard associated with working with electricity. A person who has stopped breathing is not necessarily dead but is in immediate danger. Life is dependent on oxygen, which is breathed into the lungs and then carried by the blood to every body cell. Since body cells cannot store oxygen and since the blood can hold only a limited amount (and only for a short time), death will surely result from continued lack of breathing.

However, the heart may continue to beat for some time after breathing has stopped, and the blood may still be circulated to the body cells. Since the blood will, for a short time, contain a small supply of oxygen, the body cells will not die immediately. For a very few minutes, there is some chance that the person's life may be saved.

The process by which a person who has stopped breathing can be saved is called artificial ventilation (respiration). The purpose of artificial respiration is to force air out of the lungs and into the lungs, in rhythmic alternation, until natural breathing is reestablished. Records show that seven out of ten victims of electric shock were revived when artificial respiration was started in less than three minutes. After three minutes, the chances of revival decrease rapidly.

Artificial ventilation should be given only when the breathing has stopped. Do not give artificial ventilation to any person who is breathing naturally. You should not assume that an individual who is unconscious due to electrical shock has stopped breathing. To tell if someone suffering from an electrical shock is breathing, place your hands on the person's sides at the level of the lowest ribs. If the victim is breathing, you will usually be able to feel movement.

Once it has been determined that breathing has stopped, the person nearest the victim should start the artificial ventilation without delay and send others for assistance and medical aid. The only logical, permissible delay is that required to free the victim from contact with the electricity in the quickest, safest way. This step, while it must be taken quickly, must be done with great care; otherwise, there may be two victims instead of one.

In the case of portable electric tools, lights, appliances, equipment, or portable outlet extensions, the victim should be freed from contact with the electricity by turning off the supply switch or by removing the plug from its receptacle. If the switch or receptacle cannot be quickly located, the suspected electrical device may be pulled free of the victim. Other persons arriving on the scene must be clearly warned not to touch the suspected equipment until it is deenergized.

The injured person should be pulled free of contact with stationary equipment (such as a bus bar) if the equipment cannot be quickly deenergized or if the survival of others relies on the electricity and prevents immediate shutdown of the circuits. This can be done quickly and easily by carefully applying the following procedures:

- Protecting yourself with dry insulating material.
- Using a dry board, belt, clothing, or other available nonconductive material to free the victim from electrical contact. Do NOT touch the victim until the source of electricity has been removed.

Once the victim has been removed from the electrical source, it should be determined whether the person is breathing. If the person is not breathing, a method of artificial respiration is used.

5. Cardio-Pulmonary Resuscitation

Sometimes victims of electrical shock suffer cardiac arrest or heart stoppage as well as loss of breathing.

Artificial ventilation alone is not enough in cases where the heart has stopped. A technique known as CPR has been developed to provid an aid to a person who has stopped breathing and suffered a cardiac arrest. Because you are working with electricity, the risk of electrical shock is higher than in other occupations. You should, at the earliest opportunity, take a course to learn the latest techniques used in CPR. The techniques are relatively easy to learn and are taught in courses available through the American Red Cross.

6. Lockout/Tagout

Electrical power must be removed when electrical equipment is inspected, serviced, or repaired. To ensure the safety of personnel working with the equipment, power is removed and the equipment must be locked out and tagged out.

Per OSHA standards, equipment is locked out and tagged out before any preventive maintenance or servicing is performed. Lockout is the process of removing the source of electrical power and installing a lock which prevents the power from being turned on. Tagout is the process of placing a danger tag on the source of electrical power which indicates that the equipment may not be operated until the danger tag is removed.

A danger tag has the same importance and purpose as a lock and is used alone only when a lock does not fit the disconnect device. The danger tag shall be attached at the disconnect device with a tag tie or equivalent and shall have space for the worker's name, craft, and other required information. A danger tag must withstand the elements and expected atmosphere for as long as the tag remains in place. A lockout/tagout is used when:

- Servicing electrical equipment that does not require power to be on to perform the service.
- Removing or bypassing a machine guard or other safety device.
- The possibility exists of being injured or caught in moving machinery.
- Clearing jammed equipment.
- The danger exists of being injured if equipment power is turned on.

Lockouts and tagouts do not by themselves remove power from a circuit. An approved procedure is followed when applying a lockout/tagout. Lockouts and tagouts are attached only after the equipment is turned off and tested to ensure that power is off. The lockout/tagout procedure is required for the safety of workers due to modern equipment hazards. OSHA provides a standard procedure for equipment lockout/tagout. OSHA's procedure is:

- Prepare for machinery shutdown.
- Machinery or equipment shutdown.
- Machinery or equipment isolation.
- Lockout or tagout application.
- Release of stored energy.
- Verification of isolation.

A lockout/tagout shall not be removed by any person other than the person that installed it, except in an emergency. In an emergency, the lockout/tagout may be removed only by authorized personnel. The authorized personnel shall follow approved procedures. A list of company rules and procedures are given to any person that may use a lockout/tagout. Always remember:

- Use a lockout and tagout when possible.
- Use a tagout when a lockout is impractical. A tagout is used alone only when a lock does not fit the disconnect device.
- Use a multiple lockout when individual employee lockout of equipment is impractical.
- Notify all employees affected before using a lockout/tagout.
- Remove all power sources including primary and secondary.
- Measure for voltage using a voltmeter to ensure that power is off.

7. Lockout Devices

Lockout devices are lightweight enclosures that allow the lockout of standard control devices. Lockout devices are available in various shapes and sizes that allow for the lockout of ball valves, gate valves, and electrical equipment such as plugs, disconnects, etc.

Lockout devices resist chemicals, cracking, abrasion, and temperature changes. They are available in colors to match ANSI pipe colors. Lockout devices are sized to fit standard industry control device sizes.

Locks used to lock out a device may be color coded and individually keyed. The locks are rust-resistant and are available with various size shackles.

Danger tags provide additional lockout and warning information. Various danger tags are available. Danger tags may include warnings such as "Do Not Start" "Do Not Operate" or may provide space to enter worker, date, and lockout reason information. Tag ties must be strong enough to prevent accidental removal and must be self-locking and nonreusable.

Lockout/tagout kits are also available. A lockout/tagout kit contains items required to comply with the OSHA lockout/tagout standards. Lockout/tagout kits contain reusable danger tags, tag ties, multiple lockouts, locks, magnetic signs, and information on lockout/tagout procedures. Be sure the source of electricity remains open or disconnected when returning to work whenever leaving a job for any reason or whenever the job cannot be completed the same day.

New Words and Expressions

fuse [fjuːz]	n.	保险丝，熔丝
circuit ['sɜːkɪt]	n.	电路
disconnect [ˌdɪskə'nekt]	v.	断开，切断
voltage ['vəʊltɪdʒ]	n.	[电工] 电压，伏特数
ventricular [ven'trɪkjʊlə]	adj.	心室的，膨胀的
fibrillation [ˌfaɪbrɪ'leɪʃən]	n.	[医] 心室纤维颤动
suspended [səs'pendɪd]	adj.	悬浮的
respiration [ˌrespɪ'reɪʃən]	n.	呼吸，呼吸作用
withstand [wɪð'stænd]	v.	抵挡，经受住
verification [ˌverɪfɪ'keɪʃən]	n.	确认，查证

Translation Skill

科技英语翻译技巧（九）——定语从句及同位语从句的译法（Ⅰ）

在英语的各种从句中，定语从句最为复杂，因此翻译时难度也最大。而同位语从句实际上是一种特殊的定语从句，译法与定语从句有许多相同之处，因此在此与定语从句一起讨论。

一、限制性定语从句的译法

限制性定语从句和所修饰的先行词的关系十分密切，是先行词在意义上不可缺少的修饰说明语。带有限制性定语从句的句子里，主句的含义是不完整的，要靠从句补充说明，全句概念才能表达清楚。限制性定语从句的翻译往往可以采用以下三种方法。

1. 合译法

合译法即把英语限制性定语从句译成汉语的"'的'字结构"，放在被修饰词之前，把从句和主句合译成汉语的单句。这种方法尤其适合于翻译不很长的限制性定语从句。

A molecule may be considered as the smallest particle of matter *that can exist without changing its nature*.

可以认为分子是在不改变物质性质的情况下能够存在的物质的最小微粒。

以 as 引导的限制性定语从句往往有比较固定的译法。

1）such +（名词）+ as 或 such as 常译为"像……之类，像……（这）那样的，……的一种"等。

Such liquid fuel rockets *as* are now being used for space research have to carry their own supply of oxygen.

像现在用于宇宙研究的这类液态燃料火箭，必须自己携带氧气。

2）the same...as 通常译为"和……一样的，与……相同的"。

A color transmission contains *the same* information *as* a black and white transmission.

彩色传输所容纳的信息，和黑白传输容纳的信息一样。

2. 分译法

分译法是指将定语从句与主句分开，译成并列分句。凡形式上的限制性定语从句，除简短者外，一般宜译成并列分句。限制性定语从句译成并列分句时可分为重复先行词和省略先行词两种形式。

1）译成并列分句，重复先行词的含义。有时除了重复先行词的意义之外，还可加上指示代词"这""该""其"，或把关系代词译成人称代词"它""它们"等。

A floating object displaces an amount of water *whose weight equals that of the object*.

浮体排开一定量的水，其重量等于该浮体的重量。

Gasoline is a fuel *whose vapor is readily explosive*.

汽油是一种燃料，其油汽很容易爆炸。

2）译成并列分句，省略先行词。

A fuel is a material *which will burn at a reasonable temperature and produce heat*.

燃料是一种物质，在适当温度下能够燃烧，并放出热量。

3. 溶合法

溶合法是指把原句中的主句和定语从句溶合起来，译成一个独立句子的译法。这种译法特别适用于 there be 结构带有定语从句的句型。

There are some metals which possess *the power to conduct electricity and the ability to be magnetized*.

某些金属既能导电，又能磁化。

二、非限制性定语从句的译法

非限制性定语从句主要采取分译法，作为并列分句进行汉译。

1. 分译法

非限制性定语从句只对先行词进行描写、叙述或解释，而不加以限制，很自然地多译为并列句的一个分句，甚至译成独立的简单句。

1）译成并列分句，重复先行词的含义。

A force can be shown by a straight line, *the length of which stands for the magnitude of the force*.

力可以用直线表示，其长度表示力的大小。

2）译成并列分句，省略先行词。

This type of meter is called a multimeter, *which is used to measure electricity*.

这种仪表称为万用表，用来测量电。

3）译成独立句。

Nevertheless the problem was solved successfully, *which showed that the computation was accurate*.

不过问题还是圆满地解决了，这说明计算很准确。

2. 合译法

当一些较短而且有描述性的非限制性定语从句与主句关系较密切时，也可采用合译法，即译成带"的"的前置定语。

Transistors, *which are small in size*, can make previously large and bulky radios light and small.

体积小的晶体管使得先前那种大而笨的收音机变得又轻又小。

Reading Material

Physiological Effects of Electricity

Most of us have experienced some form of electric "shock", where electricity causes our body to experience pain or *trauma*. If we are fortunate, the extent of that experience is limited to *tingles* or *jolts* of pain from *static electricity* buildup discharging through our bodies. When we are working around electric circuits capable of delivering high power to loads, electric shock becomes a much more serious issue, and pain is the least significant result of shock.

As electric current is conducted through a material, any opposition to that flow of electrons (resistance) results in a dissipation of energy, usually in the form of heat. This is the most basic and easy-to-understand effect of electricity on living tissue: Current makes it heat up. If the amount of heat generated is sufficient, the tissue may be burnt. The effect is physiologically the same as damage caused by an open flame or other high-temperature source of heat, except that electricity has the ability to burn tissue well beneath the skin of a victim, even burning internal organs.

Another effect of electric current on the body, perhaps the most significant in terms of hazard, regards the nervous system. By "nervous system" I mean the network of special cells in the body called "nerve cells" or "neurons" which process and conduct the multitude of signals responsible for regulation of many body functions. The brain, spinal cord, and sensory/motor organs in the body function together to allow it to sense, move, respond, think, and remember.

Nerve cells communicate to each other by acting as "transducers": Creating electrical signals (very small voltages and currents) in response to the input of certain chemical compounds called "neurotransmitters", and releasing neurotransmitters when stimulated by electrical signals. If electric current of sufficient magnitude is conducted through a living creature (human or otherwise), its effect will be to override the tiny electrical impulses normally generated by the neurons, overloading the nervous system and preventing both reflex and volitional signals from being able to actuate muscles. Muscles triggered by an external (shock) current will involuntarily contract, and there's nothing the victim can do about it.

This problem is especially dangerous if the victim contacts an energized conductor with his or her hands. The forearm muscles responsible for bending fingers tend to be better developed than those muscles responsible for extending fingers, and so if both sets of muscles try to contract because of an electric current conducted through the person's arm, the "bending" muscles will win, clenching the fingers into a fist. If the conductor delivering current to the victim faces the palm of his or her hand, this clenching action will force the hand to grasp the wire firmly, thus worsening the situation by securing excellent contact with the wire. The victim will be completely unable to let go of the wire.

Medically, this condition of involuntary muscle contraction is called "tetanus". Electricians familiar with this effect of electric shock often refer to an immobilized victim of electric shock as being "froze on the circuit". Shock-induced tetanus can only be interrupted by stopping the current through the victim.

Even when the current is stopped, the victim may not regain voluntary control over their muscles for a while, as the neurotransmitter chemistry has been thrown into disarray. This principle has been applied in "stun gun" devices such as Tasers, which on the principle of momentarily shocking a victim with a high-voltage pulse delivered between two electrodes. A well-placed shock has the effect of temporarily (a few minutes) immobilizing the victim.

Electric current is able to affect more than just skeletal muscles in a shock victim, however. The diaphragm muscle controlling the lungs, and the heart—which is a muscle in itself—can also be "frozen" in a state of tetanus by electric current. Even currents too low to induce tetanus are often able to scramble nerve cell signals enough that the heart cannot beat properly, sending the heart into a condition known as fibrillation. A fibrillating heart flutters rather than beats, and is ineffective at pumping blood to vital organs in the body. In any case, death from asphyxiation and/or cardiac arrest will surely result from a strong enough electric current through the body. Ironically, medical personnel use a strong jolt of electric current applied across the chest of a victim to "jump start" a fibrillating heart into a normal beating pattern.

That last detail leads us into another hazard of electric shock, this one peculiar to public power systems. Though our initial study of electric circuits will focus almost exclusively on DC (direct current, or electricity that moves in a continuous direction in a circuit), modern power systems utilize alternating current, or AC. The technical reasons for this preference of AC over DC in power systems are irrelevant to this discussion, but the special hazards of each kind of electrical power are very important to the topic of safety.

DC, because it moves with continuous motion through a conductor, has the tendency to induce muscular tetanus quite readily. AC, because it alternately reverses direction of motion, provides brief moments of opportunity for an afflicted muscle to relax between alternations. Thus, from the concern of becoming "froze on the circuit", DC is more dangerous than AC.

However, AC's alternating nature has a greater tendency to throw the heart's pacemaker neurons into a condition of fibrillation, whereas DC tends to just make the heart stand still. Once the shock current is halted, a "frozen" heart has a better chance of regaining a normal beat pattern than a fibrillating heart. This is why "defibrillating" equipment used by emergency medics works: The jolt of current supplied by the defibrillator unit is DC, which halts fibrillation and and gives the heart a chance to recover.

In either case, electric currents high enough to cause involuntary muscle action are dangerous and are to be avoided at all costs.

If at all possible, shut off the power to a circuit before performing any work on it. You must secure all sources of harmful energy before a system may be considered safe to work on. In industry, securing a circuit, device, or system in this condition is commonly known as placing it in a Zero Energy State. The focus of this unit is, of course, electrical safety. However, many of these principles apply to non-electrical systems as well.

Voltage by its very nature is a manifestation of potential energy. I even used elevated liquid as an analogy for the potential energy of voltage, having the capacity (potential) to produce current (flow), but not necessarily realizing that potential until a suitable path for flow has been established, and resistance to flow is overcome. A pair of wires with high voltage between them do not look or sound dangerous even though they harbor enough potential energy between them to push deadly amounts of current through your body. Even though that voltage isn't presently doing anything, it has the potential to, and that potential must be neutralized before it is safe to physically contact those wires.

All properly designed circuits have "disconnect" switch mechanisms for securing voltage from a circuit. Sometimes these "disconnects" serve a dual purpose of automatically opening under excessive current conditions, in which case we call them "circuit breakers". Other times, the disconnecting switches are strictly manually-operated devices with no automatic function. In either case, they are there for your protection and must be used properly. Please note that the disconnect device should be separate from the regular switch used to turn the device on and off. It is a safety switch, to be used only for securing the system in a Zero Energy State:

With the disconnect switch in the "open" position as shown (no continuity), the circuit is broken and no current will exist. There will be zero voltage across the load, and the full voltage of the source will be dropped across the open contacts of the disconnect switch. Note how there is no need for a disconnect switch in the lower conductor of the circuit. Because that side of the circuit is firmly connected to the earth (ground), it is electrically common with the earth and is best left that way. For maximum safety of personnel working on the load of this circuit, a temporary ground connection could be established on the top side of the load, to ensure that no voltage could ever be dropped across the load.

With the temporary ground connection in place, both sides of the load wiring are connected to ground, securing a Zero Energy State at the load.

Since a ground connection made on both sides of the load is electrically equivalent to short-circuiting across the load with a wire, that is another way of accomplishing the same goal of maximum safety.

Either way, both sides of the load will be electrically common to the earth, allowing for no voltage (potential energy) between either side of the load and the ground people stand on. This technique of temporarily grounding conductors in a deenergized power system is very common in maintenance work performed on high voltage power distribution systems.

A further benefit of this precaution is protection against the possibility of the disconnect switch being closed (turned "on" so that circuit continuity is established) while people are still contacting the load. The temporary wire connected across the load would create a short-circuit when the disconnect switch was closed, immediately tripping any overcurrent protection devices (circuit breakers or fuses) in the circuit, which would shut the power off again. Damage may very well be sustained by the disconnect switch if this were to happen, but the workers at the load are kept safe.

It would be good to mention at this point that overcurrent devices are not intended to provide protection against electric shock. Rather, they exist solely to protect conductors from overheating due to excessive currents. The temporary shorting wires just described would indeed cause any overcurrent devices in the circuit to "trip" if the disconnect switch were to be closed, but realize that electric shock protection is not the intended function of those devices. Their primary function would merely be leveraged for the purpose of worker protection with the shorting wire in place.

Since it is obviously important to be able to secure any disconnecting devices in the open (off) position and make sure they stay that way while work is being done on the circuit, there is need for a structured safety system to be put into place. Such a system is commonly used in industry and it is called "Lock-out/Tag-out".

A lock-out/tag-out procedure works like this: All individuals working on a secured circuit have their own personal padlock or combination lock which they set on the control lever of a disconnect device prior to working on the system. Additionally, they must fill out and sign a tag which they hang from their lock describing the nature and duration of the work they intend to perform on the system. If there are multiple sources of energy to be "locked out" (multiple disconnects, both electrical and mechanical energy sources to be secured, etc.), the worker must use as many of his or her locks as necessary to secure power from the system before work begins. This way, the system is maintained in a Zero Energy State until every last lock is removed from all the disconnect and shutoff devices, and that means every last worker gives consent by removing their own personal locks. If the decision is made to re-energize the system and one person's lock(s) still remain in place after everyone present removes theirs, the tag(s) will show who that person is and what it is they're doing.

Even with a good lock-out/tag-out safety program in place, there is still need for diligence and common sense precaution. This is especially true in industrial settings where a multitude of people may be working on a device or system at once. Some of those people might not know about proper lock-out/tag-out procedure, or might know about it but are too complacent to follow it. Don't assume that everyone has followed the safety rules!

After an electrical system has been locked out and tagged with your own personal lock, you must then double-check to see if the voltage really has been secured in a zero state. One way to check is to see if the machine (or whatever it is that's being worked on) will start up if the "Start" switch or button is actuated.

If it starts, then you know you haven't successfully secured the electrical power from it.

Additionally, you should always check for the presence of dangerous voltage with a measuring device before actually touching any conductors in the circuit. To be safest, you should follow this procedure of checking, using, and then checking your meter:

- Checking to see that your meter indicates properly on a known source of voltage.
- Using your meter to test the locked-out circuit for any dangerous voltage.
- Checking your meter once more on a known source of voltage to see that it still indicates as it should.

While this may seem excessive or even paranoid, it is a proven technique for preventing electrical shock. I once had a meter fail to indicate voltage when it should have while checking a circuit to see if it was "dead". Had I not used other means to check for the presence of voltage, I might not be alive today to write this. There's always the chance that your voltage meter will be defective just when you need it to check for a dangerous condition. Following these steps will help ensure that you're never misled into a deadly situation by a broken meter.

Finally, the electrical worker will arrive at a point in the safety check procedure where it is deemed safe to actually touch the conductor(s). Bear in mind that after all of the precautionary steps have taken, it is still possible (although very unlikely) that a dangerous voltage may be present. One final precautionary measure to take at this point is to make momentary contact with the conductor(s) with the back of the hand before grasping it or a metal tool in contact with it. Why? If, for some reason there is still voltage present between that conductor and earth ground, finger motion from the shock reaction (clenching into a fist) will break contact with the conductor. Please note that this is absolutely the last step that any electrical worker should ever take before beginning work on a power system, and should never be used as an alternative method of checking for dangerous voltage. If you ever have reason to doubt the trustworthiness of your meter, use another meter to obtain a "second opinion".

New Words and Expressions

trauma ['trɔːmə]	n.	[医] 外伤,损伤
tingle ['tɪŋgl]	n.	麻刺感
jolt [dʒəʊlt]	n.	摇晃
static electricity		静位觉,静电
neuron ['njʊərɒn]	n.	[解] 神经细胞,神经元
spinal cord		脊髓
neurotransmitter [ˌnjʊərətræns'mɪtə]	n.	神经传递素
volitional [vəʊ'lɪʃənəl]	adj.	意志的,凭意志的
actuate ['æktjʊeɪt]	v.	激励,驱使;开动
trigger ['trɪgə]	v.	引发,触发
contraction [kən'trækʃən]	n.	收缩,紧缩
tetanus ['tetənəs]	n.	[医] 破伤风
disarray [ˌdɪsə'reɪ]	v.	混乱
Taser ['teɪzə(r)]	n.	泰瑟枪(发射一束带电镖的箭使人暂时不能动弹的一种武器)

electrode [ɪˈlektrəʊd]	n.	电极
skeletal muscle		骨骼肌
diaphragm [ˈdaɪəfræm]	n.	横隔膜
scramble [ˈskræmbl]	v.	搅乱，使混杂
flutter [ˈflʌtə]	v.	悸动，乱跳
asphyxiation [ˌæsfɪksɪˈeɪʃən]	n.	窒息
cardiac arrest		心搏停止，心脏停搏
peculiar [pɪˈkjuːljə]	adj.	特别的，特殊的
DC (direct current)		[电] 直流电
AC (alternating current)		[电] 交流电
afflict [əˈflɪkt]	v.	使痛苦，折磨
pacemaker [ˈpeɪsmeɪkə]	n.	起搏器
halt [hɔːlt]	v.	使停止
defibrillate [dɪˈfɪbrɪleɪt]	v.	[医] (使) (心脏) 去纤颤，(使) (心脏) 除纤颤
harbor [ˈhɑːbə]	v.	隐匿
neutralize [ˈnjuːtrəlaɪz]	v.	便无效，中和
circuit breaker		[电工] 断路开关，断路器
continuity [ˌkɒntɪˈnjuː(ː)ɪti]	n.	连续性，连贯性
short-circuit [ˌʃɔːtˈsɜːkɪt]	v.	使短路
maintenance work		维修工作
leverage [ˈliːvərɪdʒ]	v.	用杠杆……
padlock [ˈpædlɒk]	n.	挂锁
combination lock		号码锁
control lever		操纵杆，控制杆
consent [kənˈsent]	n.	同意，允诺
common sense		常识 (尤指判断力)
complacent [kəmˈpleɪsnt]	adj.	自满的，得意的
double-check [ˈdʌbltʃek]	v.	仔细检查
meter [ˈmiːtə]	n.	仪表
paranoid [ˈpærənɔɪd]	adj.	多疑的，偏执的
precautionary [prɪˈkɔːʃənəri]	adj.	预防的
momentary contact		瞬时接触
trustworthness [ˌtrʌstˈwɜːθnɪs]	n.	可信赖，确实性

Unit Eleven

Machinery Equipment Safety

The use of machinery at work carries a number of different risks, including hazards associated with the action of the components of the machine, its integrity or its operation. Effective machinery safety can be achieved through better design, improved layout and management procedures, including the selection and training of operators.

Many accidents in the workplace arise out of the use of machinery and tools ranging from hand tools through machine tools such as lathes and presses to large items of mobile mechanical plant e. g. cranes or fork lift trucks. Injuries may be caused in many ways such as:

- Articles dropped or falling, e. g. from lifting plant.
- Contact with moving parts of machinery.
- Trapping, particularly of hands.
- Electric shocks, burns or scalds.
- Flying debris following fragmentation of machinery or materials being worked on.
- Emission of harmful substances from machinery.

Accidents not only cause human suffering, they also cost money, for example in lost working hours, training temporary staff, insurance premiums, fines and managers' time. By using safe, well-maintained equipment operated by adequately trained staff, you can help prevent accidents and reduce these personal and financial costs.

"Work equipment" is almost any equipment used by a worker at work including: Machines such as circular saws, drilling machines, photocopiers, mowing machines, tractors, dumper trucks and power presses; hand tools such as screwdrivers, knives, hand saws and meat cleavers; lifting equipment such as lift trucks, elevating work platforms, vehicle hoists, lifting slings and bath lifts; other equipment such as ladders and water pressure cleaners.

1. What Risks Are There from Using Work Equipment?

Many things can cause a risk, for example:

- Using the wrong equipment for the job, e. g. ladders instead of access towers for an extended job at high level.
- Not fitting adequate guards on machines, leading to accidents caused by entanglement, shearing, crushing, trapping or cutting.
- Not fitting adequate controls, or the wrong type of controls, so that equipment cannot be turned off quickly and safely, or starts accidentally.
- Not properly maintaining guards, safety devices, controls, etc. so that machines or equipment become unsafe.
- Not providing the right information, instruction and training for those using the equipment.
- Not fitting roll-over protective structures (ROPS) and seat belts on mobile work equipment where there is a risk of roll over (Note: this does not apply to quad bikes).
- Not maintaining work equipment or carrying out regular inspections and thorough examinations.
- Not providing the personal protective equipment needed to use certain machines safely, e. g. chainsaws, angle grinders.

2. How to Reduce the Risks

(1) **Use the Right Equipment for the Job**

Many accidents happen because people have not chosen the right equipment for the work to be done. Controlling the risk often means planning ahead and ensuring that suitable equipment or machinery is available.

(2) **Make Sure Machinery Is Safe**

You should check the machinery is suitable for the work—think about how and where it will be used. All new machinery should be:
- CE marked.
- Safe—never rely exclusively on the CE mark to guarantee machinery is safe. It is only a claim by the manufacturer that the equipment is safe. You must make your own safety checks.
- Provided with an EC Declaration of Conformity (ask for a copy if you have not been given one).
- Provided with instructions in English.

If you think that machinery you have bought is not safe, DO NOT USE IT. Contact the manufacturer to discuss your concerns and if they are not helpful, contact your local HSE office for advice. Remember, it is your responsibility as an employer or a subcontractor to ensure you do not expose others to risk.

(3) **Guard Dangerous Parts of Machines**

Controlling the risk often means guarding the parts of machines and equipment that could cause injury. Remember: Use fixed guards wherever possible, properly fastened in place with screws or nuts and bolts which need tools to remove them; if employees need regular access to parts of the machine and a fixed guard is not possible, use an interlocked guard for those parts. This will ensure that the machine cannot start before the guard is closed and will stop if the guard is opened while the machine is operating; in some cases, e. g. on guillotines, devices such as photoelectric systems or automatic guards may be used instead of fixed or interlocked guards; check that guards are convenient to use and not easy to defeat, otherwise they may need modifying; make sure the guards allow the machine to be cleaned and maintained safely; where guards cannot give full protection, use jigs, holders, push sticks etc. to move the workpiece.

(4) Make Sure Machinery and Equipment Are Maintained in a Safe Condition

To control the risk you should carry out regular maintenance and preventive checks, and inspections where there is a significant risk. Some types of equipment are also required by law to be thoroughly examined by a competent person. Inspections should be carried out by a competent person at regular intervals to make sure the equipment is safe to operate. The intervals between inspections will depend on the type of equipment, how often it is used and environmental conditions. Inspections should always be carried out before the equipment is used for the first time or after major repairs. Keep a record of inspections made as this can provide useful information for maintenance workers planning maintenance activities.

- Make sure the guards and other safety devices (e.g. photoelectric systems) are routinely checked and kept in working order. They should also be checked after any repairs or modifications by a competent person.
- Check what the manufacturer's instructions say about maintenance to ensure it is carried out where necessary and to the correct standard.
- Routine daily and weekly checks may be necessary, e.g. fluid levels, pressures, brake function, and guards. When you enter a contract to hire equipment, particularly a long-term one, you will need to discuss what routine maintenance is needed and who will carry it out.
- Some equipment, e.g. a crane, needs preventive maintenance (servicing) so that it does not become unsafe.
- Lifting equipment, pressure systems and power presses should be thoroughly examined by a competent person at regular intervals specified in law or according to an examination scheme drawn up by a competent person. Your insurance company may be able to advise on who would be suitable to give you this help.

(5) Carry out Maintenance Work Safely

Many accidents occur during maintenance work. Controlling the risk means following safe working practices, for example:

- Where possible, carry out maintenance with the power to the equipment off and ideally disconnected or with the fuses or keys removed, particularly where access to dangerous parts will be needed.
- Isolate equipment and pipelines containing pressurized fluid, gas, steam or hazardous material. Isolating valves should be locked off and the system depressurized where possible, particularly if access to dangerous parts will be needed.
- Support parts of equipment which could fall.
- Allow moving equipment to stop.
- Allow components which operate at high temperatures time to cool.
- Switch off the engine of mobile equipment, put the gearbox in neutral, apply the brake and, where necessary, chock the wheels.
- To prevent fire and explosions, thoroughly clean vessels that have contained flammable solids, liquids, gases or dusts and check them before hot work is carried out. Even small amounts of flammable material can give off enough vapour to create an explosive air mixture which could be ignited by a hand lamp or cutting/welding torch.
- Where maintenance work has to be carried out at height, ensure that a safe and secure means of access is provided which is suitable for the nature, duration and frequency of the task.

3. The Precaution Means in Practice

Accidents using the following equipment are common in small firms, but they can be prevented by some simple rules.

(1) **Drilling Machines**

As with other cutting machines, the operator must be protected from the rotating chuck and swarf that is produced by the drill bit. Specially designed shields can be attached to the quill and used to protect this area. A telescoping portion of the shield can retract as the drill bit comes down into the workpiece. On larger gang or radial drills, a more universal type shield is typically applied.

(2) **Lathes**

There are three main mechanical safety considerations for lathes (engine, turret, etc.). One is the rotating chuck that could catch the operator's clothing, jewelry, hair, or hand and pull them into the machine. Two is the hazardous flying chips and coolant splash that are generated at the point of operation (where the tool contacts the workpiece being machined). The last safety consideration is a chuck wrench left in the chuck. To protect these areas, two shields can be applied—one around a portion of the chuck and the other at the point of operation. Larger sliding shields can protect both areas, providing the workpiece is not too long. On VTLs (vertical turret lathes), the safety concern is the rotating table and the point-of-operation swarf. Special barriers may have to be fabricated around the tables of these machines; shields can be provided at the point of operation.

(3) **Milling Machines**

The main mechanical safety consideration for milling machines is the swarf that is generated at the point of operation. Another safety concern is the tool cutter, which could catch operator's clothing, jewelry, hair, or any other part of the body. Usually on smaller mills, the operator and other employees in the machine area are protected by shields. These shields can be applied around the perimeter of the table or bed area or close to the cutter, depending on the size of the workpiece and the application. On larger milling machines, operators are sometimes protected by location; however, when working close to a cutting tool they must be protected from swarf.

(4) **Grinding Machines**

Shields are usually applied to grinding machines to protect the operator from chips (swarf), sparks, coolant, or lubricant. A vacuum pedestal is also available to capture discharged debris. Other safety concerns for grinders are the adjustment of the work rests and the adjustable tongues or ends of the peripheral members at the top of each wheel. Work rests shall be kept adjusted closely to the wheel with a maximum opening of 1/8 in. The distance between the wheel periphery and the adjustable tongue or the end of the peripheral member at the top shall never exceed 1/4 in.

New Words and Expressions

lathe [leɪð]	n.	车床
crane [kreɪn]	n.	起重机
fork lift truck		叉车
insurance premium		保险费

circular saw	圆锯
photocopier [ˈfəʊtəʊkɒpɪə]	n. 复印机
mowing machine	割草机
screwdriver [ˈskruːdraɪvə]	n. 螺钉旋具
hand saw	手锯
vehicle hoist	升车机
lifting sling	升降索套
entanglement [ɪnˈtæŋglmənt]	n. 纠缠，缠结
safety device	安全防护装置
ROPS (roll-over protective structure)	翻车安全保护装置
angle grinder	角度磨床
CE marked	CE (Conformite Europeenne) 认证
Declaration of Conformity	符合性声明
HSE	健康 (Health)、安全 (Safety) 和环境 (Environment) 管理体系的简称
guillotine [gɪləˈtiːn]	n. 剪床
isolating valve	隔离阀
drilling machine	钻孔机（钻床）
swarf [swɔːf]	n. 切屑
VTL (vertical turret lathe)	立式转塔车床
milling machine	铣床
workpiece [ˈwɜːkpiːs]	n. 工件
grinding machine	磨床
coolant [ˈkuːlənt]	n. 冷冻剂
lubricant [ˈluːbrɪkənt]	n. 润滑物，润滑油，润滑剂
vacuum [ˈvækjʊəm]	n. 真空，空间
vacuum pedestal	真空助力器座

Translation Skill

科技英语翻译技巧（十）——定语从句及同位语从句的译法（II）

一、带有状语意义的定语从句的译法

有的定语从句跟主句在逻辑上有状语关系，说明原因、结果、目的、时间、条件或让步等。

1. 译成表示原因的分句

Aluminum copper alloy *which when heat-treated has good strength at high temperature* is used for making pistons and cylinder heads for automobiles.

由于铜铝合金经过热处理后具有良好的高温强度，所以用来制造汽车发动机的活塞与气缸盖。

2. 译成表示结果的分句

Copper, *which is widely used for carrying electricity*, offers very little resistance.

铜的电阻很小,所以广泛地用来传输电力。

3. 译成表示让步的分句

Friction, *which is often considered as a trouble*, is sometimes a help in the operation of machines.

摩擦虽然常被认为是一种麻烦,但有时却有助于机器的运转。

4. 译成表示条件的分句

For any machine *whose input and output forces are known*, its mechanical advantage can be calculated.

对于任何机器来说,如果知其输入力和输出力,就能求出其机械效益。

5. 译成表示目的的分句

An improved design of such a large tower must be achieved *which results in more uniformed temperature distribution in it*.

这种大型塔的设计必须改进,以保证塔内温度分布更为均匀。

6. 译成表示时间的分句

Electricity *which is passed through the thin tungsten wire inside the bulb* makes the wire very hot.

当电通过灯泡里的细钨丝时,会使钨丝达到很高的温度。

二、特种定语从句的译法

所谓特种定语从句,是指修饰整个主句或主句部分内容的非限制性定语从句。这种定语从句只能由 which 和 as 引导。

1. 由 which 引导的特种定语从句

这种从句总是位于主句之后,通常说明整个主句,其前有逗号分开。一般采用分译法,which 常译成"这",有时也译成"从而,因而"等。

The sun heats the earth, *which makes it possible for plants to grow*.

太阳供给地球热能,这使植物生长有了可能。

To find the pressure we divide the force by the area on which it presses, *which gives us the force per unit area*.

欲求得压强,需把力除以它所作用的面积,从而得出单位面积上的压力。

2. 由 as 引导的特种定语从句

在这种定语从句中,as 通常指整个主句的内容或主句的部分内容,其位置十分灵活,不仅可以位于主句之后,还可以位于主句之前,有时也可能位于主句当中,相当于表示说话者态度或看法的插入语。翻译时主句与从句分译,往往把 as 译成"正如……那样,这,如,像"等。

As is mentioned above, the function of the device is wonderful.

如上所述,这种装置的功能十分出色。

These two pipes are not properly aligned, *as you can see from that position*.

这两根管子没有完全对直,这可以从那个位置上看出来。

三、定语从句与先行词之间的分隔与译法

定语从句有时会被其他成分隔开,翻译这种定语从句关键在于正确理解,通过语法现象和逻辑判断确定定语从句所修饰的对象。翻译时总的原则与前述定语从句的译法相同。

1. 定语从句与先行词之间被定语短语或状语分隔

The concept of energy leads to the principle of the conservation of energy, *which unifies a wide range of phenomena in the physical science*.

对能量的理解导致了能量守恒定律，该定律统一了物理科学中相当众多的现象。

2. 定语从句与先行词之间被谓语分隔

The day will come *when coal and oil will be used as raw materials rather than as fuels*.

煤和石油用作原料而不是燃料的日子一定会到来。

四、同位语从句的译法

由于同位语从句在作用上很接近定语从句，因此同位语从句的译法与定语从句基本相同，可采用合译法、分译法和转译法。

1. 合译法

合译法即把同位语从句译成前置定语，一般在所说明的名词前加个"的"字，有时还可添加"这种""这一""这个""那个"等词，放在同位语从句后面作为同位成分。

The fact *that the gravity of the earth pulls everything towards the center of the earth* explains many things.

地球引力把一切东西都吸向地心这一事实解释了许多现象。

2. 分译法

采用分译法，即把同位语从句译成独立句子时，往往要在它的前面加冒号、破折号或"即"字。这种方法尤其适用于较长的同位语从句。

We have come to the correct conclusion *that two volumes of hydrogen and one volume of oxygen are united into one volume of steam*.

我们已得出正确的结论：两个体积的氢和一个体积的氧化合成一个体积的水蒸气。

3. 转译法

如果被同位语从句说明的本位语是含有动作意义的名词，如 hope，knowledge，assurance 等，一般可把这类名词译为动词，而将同位语从句译为汉语的主谓词组或动宾词组，做该动词的宾语。

Even the most precisely conducted experiments offer no hope *that the results can be obtained without any error*.

即使进行的是最精确的实验，也没有希望获得无任何误差的实验结果。

Machine Guarding

Machine hazards are sources of potential harm or a situation of potential harm. The hazards can cause human injuries or ill-health as well as damage to property or the environment. The concept of hazard can also be divided into hazards and hazardous situations. Hazards are sources of possible injuries or damage to health and hazardous situations are any situations in which a person is exposed to hazards.

Hazards create potential conditions waiting to become loss. Unplanned events that cause losses are called accidents. An accident is a dynamic mechanism that begins with the activation of a hazard, flows through the system as a series of events in logical sequences and finally produces a loss. If the unplanned event has a potential to lead to accident, it is incident.

The concept of risk is essential in estimating and evaluating the significance of the losses. The risk describes the potential for realization of unwanted and negative consequences of events. In machine design the risk is described as a combination of the probability and the degree of the possible injury or damage to health in a hazardous situation.

Machines must be designed according to the essential safety and health requirements on the basis of the state of the art and a manufacturer must continuously follow the technological possibilities that can be applied to improve machine safety. Safety is machine's ability to perform its function without causing injury or damage to health. The machine is safe if the risks of the machine are judged to be acceptable.

Many serious accidents at work involve machinery and occur for several reasons, including:
- Badly designed machine guards (e. g. those that can be removed).
- Poor maintenance of machines and guards.
- Guards are not provided.
- No supervisory system to ensure that guards are used.
- Payment/bonus systems that encourage "shortcuts" (e. g. guards can be removed if they restrict production).
- Lack of adequate training for workers on the safe use of machines.

All too often workers can get caught in unguarded machinery. A combination of unguarded machines and loose clothing, long hair, dangling chains, gloves, rings, etc., can be fatal. Machine guards are essential for protecting workers from needless and preventable injuries. A good rule of thumb to remember is that any machine part, function, or process which may cause injury must be guarded. Where the operation of a machine, or accidental contact with it, can injure the operator or other workers in the immediate area, the machine must be guarded.

1. Core Information

How do workers get injured at machines? Basically a worker may be injured at machinery as a result of:
- Coming in contact with moving parts of a machine—being hit or getting caught.
- Getting trapped between moving parts of a machine or material and any fixed structure.
- Being hit by material or parts which have been thrown out of the machine.

Where do mechanical hazards occur? There are three basic parts of a machine that must be guarded:
- The point of operation—the actual point where the work is performed on the material, such as cutting or sewing.
- The power transmission apparatus—all components of the mechanical system that transmit energy to the part of the machine performing the work. These components include flywheels, pulleys, belts, connecting rods, couplings, cams, spindles, chains, cranks, and gears.
- Other moving parts—all parts of the machine which move while the machine is working. These can include reciprocating, rotating, and transverse moving parts, as well as feeder mechanisms and auxiliary parts of the machine.

There are a number of mechanical motions and actions that can be hazardous to workers if safeguards are not present. Most machines perform their function of cutting, shearing, bending or punching by a series of mechanical motions, namely through rotation of machine parts (produces nip point hazards); through reciprocation of machine parts (this basically refers to up-and-down or back-and-forth motion as with sewing machines); and, transverse motion (movement in a straight, continuous line) in which a worker may be struck or caught in a pinch point by the moving part.

2. Requirements for, and Types of, Machine Guards

There are a number of general requirements for all machine guards if they are to protect workers against mechanical hazards. These include:

- Prevent contact—the guard must prevent hands, arms, or any other part of a worker's body from coming in contact with dangerous moving parts. A good guard eliminates the possibility of the operator placing their hands near dangerous moving parts.
- Secure—the guards should be firmly fixed to the machine (or preferably an integral part of the machine) and not easily removed. Guards should be made of durable materials that can withstand workplace conditions over the lifetime of the machine.
- Protect from falling objects—the guard should ensure that no objects (such as a tool) can fall into moving parts.
- Create no new hazards—a guard defeats its own purpose if it creates another hazard such as its own shear point, a jagged or sharp edge, which can cause cuts.
- Create no interference with work—any guard which impedes a worker from performing his/her job efficiently and in comfort, is likely to be removed or overridden. It should be pointed out that proper guards on machines can actually improve productivity as workers then have confidence that they will not be injured.
- Allow safe lubrication—if possible, the operator or maintenance worker should be able to lubricate the machine safely without removing the guard or having to reach inside the machine and into any hazardous area.

There are various types of guards that can be used to prevent injury in the workplace, including:

- Fixed guards—these are the most common type of guard found in garment factories and basically prevent any contact between hands, arms, etc. and any moving machine parts. They should not easily be removed. In the best cases they provide the maximum amount of protection and require the minimum amount of maintenance. In some cases these guards can interfere with visibility.
- Interlocking guards—when this type of guard is opened or removed, the tripping mechanism and/or power automatically shuts off or disengages. The machine cannot start until the guard is back in place.
- Adjustable guards—these guards are useful because, as the name implies, they allow flexibility in accommodating various sizes (thickness, width, height, etc.) of stock. The best examples in a garment factory are the band knives in the cutting section where the guards are adjustable and can deal with varying thicknesses of material blocks.
- Self-adjusting guards—the openings of these barriers are determined by the movement of the stock. As the operator moves the stock into the danger area, the guard is pushed away, providing an opening which is only large enough to admit the stock. After the stock is removed, the guard returns to the rest position. In some cases, these guards can interfere with visibility.

There are also a number of other techniques that can be used including—photoelectric trips, restraints, two-hand controls, safety trips, pullback mechanisms, and automatic feeds. Their name describes the mode of action but all work on a principle of disabling the machine as soon as a part of the body gets near a dangerous, moving component.

3. Some of the Main Machine Guarding Problems in a Garment Factory

The most common accidents in the garment industry tend to be:
- Cut fingers of machine operators in the cutting section.
- A needle in the finger of sewers.
- Burns from irons in the ironing section.

As with all machinery, the machines in the garment industry are potentially dangerous. The most obvious are probably the band knives in the cutting section. If you take a walk-through survey of a garment factory, one of the most common observations is the number of machines that have guards missing or that the guards are inadequate. As has been discussed, it is important that you purchase so-called safe machines with fixed guards as an integral part—guards that cannot be removed, that allow for safe maintenance, and that provide clear visibility.

4. Accidents and Unguarded Machines

Tragically, accidents occur in all workplaces and the garment industry is no different. Accidents are costly to the employer in terms of loss of productivity; to the worker in terms of injury, loss of wages, etc.; and to society as a whole. The three factors that contribute to accidents are:
- faulty technical equipment;
- the working environment;
- the worker.

All machines in the garment industry can be dangerous. Machine guards have been developed to prevent accidents happening when workers get too close to dangerous machine parts. These parts should either be out of reach or enclosed so that no one can touch them or fall into them accidentally. It is best to purchase machines with guards included as an integral part.

All guards must be robust and cannot be removed easily. All workers should be trained in the safe use of machinery and shown how to stop machines in the case of an emergency. Workers must wear the correct personal protective equipment (PPE) where appropriate and should not wear dangling chains or necklaces, loose clothing, gloves, rings or long hair which could get caught up in moving parts.

New Words and Expressions

dynamic [daɪˈnæmɪk]   adj. 动态的，有动力的，有力的
activation [æktɪˈveɪʃən]   n. 激活
supervisory [sjuːpəˈvaɪzəri]   adj. 管理的，监控的
bonus [ˈbəʊnəs]   n. 红利，奖金
shortcut [ˈʃɔːtkʌt]   n. 捷径

coupling ['kʌplɪŋ]   n. 联结，结合，耦合
reciprocating [rɪ'sɪprəkeɪtɪŋ]   adj. 往复的，来回的
safeguard ['seɪfgɑːd]   n. 保卫，防护措施、设备
interlocking [ɪntə(ː)'lɒkɪŋ]   adj. 连锁的
adjustable [ə'dʒʌstəb(ə)l]   adj. 可调整的
photoelectric [fəutəuɪ'lektrɪk]   adj. 光电的
emergency [ɪ'mɜːdʒnsi]   n. 紧急情况，突发事件
personal protective equipment（PPE）   个人防护设施

Unit Twelve

Accident Analysis in Construction

In construction work many of the hazards are obvious. Most of them can be found on almost every site. The causes of accidents are well known and often repeated. Too often hazards are just seen as an inevitable part of the job, so no action is taken to control the risks they create. Consequently, the rate of accidents and ill health remains high. Action is needed to change this. This section identifies the most common causes of death, injury and ill health and sets out straightforward precautions. Applying this advice will make work safer and, in most cases, improve efficiency. Some activities (e. g. roof work and steel erection) are considered in detail, but in general most operations will present a number of hazards which are dealt with on a number of pages.

The construction industry is economically important as it typically contributes 10% of a developing country's GNP. It is also very hazardous with almost six times as many fatalities and twice as many injuries per hour worked relative to a manufacturing industry. A researcher analyses 739 construction fatalities that occurred in the UK. He found that 52% occurred due to falls from roofs, scaffolds and ladders. Falling objects and material were involved in 19.4% of the deaths, and transportation equipment (e. g. excavators and dumpers) were involved in 18.5%. He also found that 5% of construction accidents occur during excavation work.

The categories used for classifying fatal accidents were:
- falls;
- falling material and objects;
- electrical hazards;
- transport and mobile plants;
- other.

The majority of accidents that involved falls occur during work on roofs, scaffolds and ladders. Collapses of structures and falling materials also account for a large proportion of fatalities. Many of the safety hazards are specific to the different trades, and typically construction workers underestimate the hazards in their own work which affects the motivation for adopting safe work procedures. The establishment and use of procedures and regulations to enhance safety can avoid a large proportion of these accidents. There are also forceful monetary incentives in construction safety as it is estimated that construction accidents amount to about 6% of total building costs; this should encourage the industry as well as the regulatory agencies to invest in construction safety.

An article categorises the primary events as follows:
- fall of person;
- overexertion or strenuous movement;
- handling accidents;
- struck by falling or flying objects;
- contact with stationary objects, missed steps, etc;
- contact with moving objects;
- contact with heat/cold;
- contact with chemicals;
- exposure to or contact with electricity;
- fire;
- explosion or blast;
- unclassified.

Based on a study of cases from UK, USA, France, Canada and Sweden, the author also provided a comparison of construction and manufacturing for injury incidence rates both for different parts of the body and categories of the workers. Supervisors experienced a very high rate of falls injuries and the highest rate of accidents from stepping on objects. For plumbers the highest rate of accidents involved being "struck by falling or flying objects" and for masons the highest rate was for "overexertion or strenuous movement". The five most dangerous hand tools for all categories of workers were:
- knife;
- hammer, sledge hammer, etc;
- grinding, cutting machine;
- jackhammer;
- drill.

1. Working at Height

Falls are the largest cause of accidental death in the construction industry. They account for 50% of those accidentally killed. Most accidents involving falls could have been prevented if the right equipment had been provided and properly used. All falls need to be prevented. However, specific precautions need to be taken (guard rails, barriers, etc.) where it is possible to fall 2m or more. When planning for work at height, consider where the work will be done. Obviously the first choice will be any existing structure which allows safe access and provides a safe working place. Where it is not possible to work safely from the existing structure, an extra working platform will be needed.

Suitable precautions should be taken to prevent falls. Guard rails, toe boards and other similar barriers should be provided whenever someone could fall 2m or more. They should be made from any material providing they are strong and rigid enough to prevent people from falling and be able to withstand other loads likely to be placed on them. For example, guard rails fitted with brick guards need to be capable of supporting the weight of stacks of bricks which could fall against them.

Barriers other than guard rails and toe boards can be used, so long as they are at least 910 mm high, secure and provide an equivalent standard of protection against falls and materials rolling, or being kicked, from any edges. If the risk comes from falling through openings or fragile material (for example, rooflights or asbestos roof sheets), an alternative to guard rails or a barrier is to cover the opening or material.

Safe working platforms—Working platforms are the parts of structures upon which people stand while working. As well as being adequately supported and provided with guard rails or barriers. Working platforms should be constructed to prevent materials from falling. As well as toe boards or similar protection at the edge of the platform, the platform itself should be constructed to prevent any object which may be used on the platform from falling through gaps or holes, causing injury to people working below. For scaffolds, a close-boarded platform would suffice, although for work over public areas, a double-boarded platform sandwiching a polythene sheet may be needed. If cradles are used and they have meshed platform floors, the mesh should be fine enough to prevent materials, especially nails and bolts, from slipping through.

(1) **General Access Scaffolds**

For any scaffold make sure: It is based on a firm, level foundation. The ground or foundation should be capable of supporting the weight of the scaffold and any loads likely to be placed on it. Watch out for voids such as basements or drains, or patches of soft ground, which could collapse when loaded. Provide extra support as necessary.

(2) **Mobile Elevating Work Platforms** (MEWPs)

Some MEWPs are described as suitable for rough terrain. This usually means that they are safe to use on some uneven or undulating ground—but check their limitations in the manufacturer's handbook before taking them onto unprepared or sloping ground. Wearing a harness with a lanyard attached to the platform can provide extra protection against falls, especially while the platform is in motion.

(3) **Roof Work**

Almost one in five workers killed in construction accidents are doing roof work. Most of these are specialist roofers, but some are simply involved in maintaining and cleaning roofs. Some of these workers die after falling off the edges of flat and sloping roofs. Many other workers die after falling through fragile materials. Many roof sheets and rooflights are, or can become, fragile. Asbestos cement, fiberglass and plastic generally become more fragile with age. Steel sheets may rust. Sheets on poorly repaired roofs might not be properly supported.

2. Excavations

Every year people are killed or seriously injured while working in excavations. Many are killed or injured by collapses and falling materials, some are killed or injured when they contact buried underground services. Groundwork has to be properly planned and carried out to prevent accidents. Before digging any trenches, pits, tunnels, or other excavations, decide what temporary support will be required and plan the precautions that are going to be taken against:

Collapse of the sides or roof—Prevent the sides from collapsing by battering them to a safe angle or supporting them with sheeting or proprietary support systems. Take similar precautions to prevent the face from collapsing. Install support without delay as the excavation progresses. Never work ahead of the support.

The work should be directed by a competent supervisor. Give the workers clear instructions.

Materials falling into excavations—Do not store excavated spoil and other materials or park plant and vehicles close to the sides of excavations. The extra loadings from spoil, vehicles, etc. can make the sides of excavations more likely to collapse. Loose materials may fall from spoil heaps, etc. into the excavation. A scaffold board used as a toe board and fixed along the outside of the trench sheets will provide extra protection against loose materials falling. Hard hats will protect those working in the excavation from small pieces of materials falling either from above, or from the sides of the excavation.

People and vehicles falling into excavations—Prevent people from falling by guarding excavations. Edges of excavations more than 2m deep should be protected with substantial barriers where people are liable to fall into them. All excavations in public places should be suitably fenced off to prevent members of the public approaching them.

Prevent vehicles from falling into excavations by keeping them out of the area. Vehicles passing close to the edges of excavations may also overload the sides, leading to collapse. Where necessary, use baulks or barriers to keep vehicles away from excavated edges. Baulks and barriers are best painted or marked to make sure they can be seen by drivers.

3. Working in Confined Spaces

Not knowing the dangers of confined spaces has led to the deaths of many workers. Often those killed include not only those working in the confined space but also those who, not properly equipped, try to rescue them. Work in such spaces requires skilled and trained people to ensure safety. If work cannot be avoided in a confined space, it will often be safer to bring in a specialist for the job.

Air in the confined space is made unbreathable either by poisonous gases and fumes or by lack of oxygen. There is not enough natural ventilation to keep the air fit to breathe. In some cases the gases may be flammable, so there may also be a fire or explosion risk. Working space may be restricted, bringing workers into close contact with other hazards such as moving machinery, electricity or steam vents and pipes. The entrance to a confined space, for example, a manhole, may make escape or rescue in an emergency more difficult.

4. Moving, Lifting and Handling Loads

Many construction workers are killed or seriously injured during lifting operations because of accidents such as: cranes overturning, material falling from hoists and gin wheels collapsing. Many more suffer long-term injury because they regularly lift or carry items which are heavy or awkward to handle, for example: lifting dense concrete blocks, paviours laying slabs and labourers lifting and carrying bagged products, such as cement and aggregates.

5. Site Vehicles and Mobile Plant

Workers are killed every year on construction sites by moving vehicles or vehicles overturning. Many more are seriously injured in this way. The risks can be reduced if the use of vehicles and mobile plant is properly managed.

6. Electricity

Electrical equipment is used on virtually every site. Everyone is familiar with it, but not all seem to remember that electricity can kill. Electrical systems and equipment must be properly selected, installed, used and maintained.

New Words and Expressions

GNP (Gross National Product)	国民生产总值
manufacturing industry	制造业
roof [ruːf]	n. 屋顶，房顶，顶
scaffold [ˈskæfəld]	n. 脚手架
ladder [ˈlædə]	n. 梯子，阶梯
transportation equipment	运输设备
excavator [ˈekskəveɪtə]	n. 挖掘机，凿岩机
dumper [ˈdʌmpə]	n. 卸货车，倾卸装置
excavation work	开挖施工
category [ˈkætɪɡəri]	n. 种类，范畴
handle [ˈhændl]	v. 搬运
strike [straɪk]	v. 撞击，打击
plumber [ˈpʌmbə]	n. 管道工
mason [ˈmeɪsn]	n. 泥瓦匠
sledge hammer	大锤
grind [ɡraɪnd]	v. 粉碎
cutting machine	切割机
jackhammer [ˈdʒækˌhæmə(r)n]	n. 手持式风钻，风镐
guard rail	防护围栏
toe board	趾板
rooflight sheet	采光屋面板
asbestos [æzˈbestɒs]	n. 石棉
polythene [ˈpɒlɪθiːn]	n. 聚乙烯
cradle [ˈkreɪdl]	n. 托板
terrain [ˈtereɪn]	n. 地形
fiberglass [ˈfaɪbəɡlɑːs]	n. 玻璃纤维
confined space	狭小空间
manhole [ˈmænˌhəʊl]	n. 检修孔
paviour [ˈpeɪvjə]	n. 铺设工人
slab [slæb]	n. 厚平板，混凝土路面

Translation Skill

科技英语翻译技巧（十一）——状语从句的译法

在状语从句的翻译中，主要应注意状语从句的位置、连词的译法和省略以及状语从句的转译等。

英译汉时状语从句的位置尤其重要。一般来说，汉语中，状语从句多半在主句前面，有时放在整个句子当中。此外，汉译时连词常可省略，这在时间、条件状语从句中尤为常见。另外，不要碰到 when 就译成"当"，碰到 if 就译成"如果"，碰到 because 就译成"因为"，而应酌情进行变化或简化。

The computer will find the route *when* you send your signal to it.

把信号输入计算机，它就会找到行车路线。

If water is cold enough, it changes to ice.

水温降到一定程度便会结冰。

另外，科技英语翻译中，有时将时间状语从句转译成条件状语从句，将地点状语从句转译成条件状语从句。

These three colors, red, green, and blue, *then combined*, produce black.

红、绿、蓝三色混在一起，就会变成黑色。

Where there is nothing in the path of the beam of light, nothing is seen.

如果光轨迹上没有东西，就什么也看不出来。

一、时间状语从句的译法

1. 译成相应的时间状语，放在句首

不论原文中表示时间的从句是前置或后置，根据汉语习惯，都要译在其主句的前面。

Heat is always given out by one substance and taken in by another *when heat-exchange takes place*.

热交换发生时，总是某一物质释放热量，另一物质吸收热量。

2. 译成并列句

有的连词（如 as, while, when 等）引导时间状语从句，在表达主句和从句的谓语动作同时进行时，英译汉时可省略连词，译成汉语的并列句。

The earth turns round its axis *as it travels around the sun*.

地球一面绕太阳运行，一面绕地轴回转。

3. 译成条件状语从句

when 等引导的状语从句，若从逻辑上判断具有条件状语的意义，则往往可转译成条件状语从句。

Turn off the switch *when anything goes wrong with the machine*.

如果机器发生故障，就把开关关上。

二、地点状语从句的译法

1. 译成相应的地点状语

一般可将地点状语从句译在句首。

Heat is always being transferred in one way or another, *where there is any difference in temperature*.

凡有温差的地方，热都会以这样或那样的方式传输。

2. 译成条件状语从句或结果状语从句

where 或 wherever 引导的状语从句，若从逻辑上判断具有条件状语或结果状语的意义，则可转译为相应的状语从句。

Where water resources are plentiful, hydroelectric power stations are being built in large numbers.

只要是水源充足的地方，就可以修建大批的电站。

It is hoped that solar energy will find wide application *wherever it becomes available.*

可以期待，太阳能将得到广泛的利用，以至于任何地方都可以使用。

三、原因状语从句的译法

1. 译成表"因"的分句

一般来说，汉语表"因"的分句置于句首，英语则较灵活。但现代汉语中，也有放在后面的情况，此时往往含有补充说明的意义。

Some sulphur dioxide is liberated when coal, heavy oil and gas burn, *because they all contain sulphur compounds.*

因为煤、重油和煤气都含有硫化物，所以它们燃烧时会放出一些氧化硫。

To launch a space vehicle into orbit, a very big push is needed *because the friction of air and the force of gravity are working against it.*

要把宇宙飞行器送入轨道，需要施加很大推力，因为空气的摩擦力和地球引力对它起阻碍作用。

2. 译成因果偏正复句的主句

这实际是一种省略连词的译法，把从句译成主句。

Because energy can be changed from one form into another, electricity can be changed into heat energy, mechanical energy, light energy, etc.

能量能从一种形式转换成另一种形式，所以电可以转变为热能、机械能、光能等。

四、条件从句的译法

1. 译成表示"条件"或"假设"的分句

按照汉语的习惯，不管表示条件还是假设，分句都放在主句的前部，因此英语的条件从句汉译时绝大多数置于句首。

Unless you know the length of one side of a square or a cube, you cannot find out the square's area or the cube's volume.

除非已知一个正方形或一个正方体的边长，否则就无法求出这个正方形的面积或这个正方体的体积。

2. 译成补充说明情况的分句

不少条件状语从句汉译时可置于主句后面，做补充说明情况的分句。

Iron or steel parts will rust, *if they are unprotected.*

铁件或钢件是会生锈的，如果不加保护的话。

五、让步状语从句

1. 译成表示"让步"的分句

汉语中让步分句一般前置，但也可后置。

Though we get only a relatively small part of the total power radiated from the sun, what we get is much more than enough for our needs.

虽然我们仅得到太阳辐射总能量的一小部分,但是,与我们的实际需要量相比,这已绰绰有余了。

Energy can neither be created nor destroyed *although its form can be changed*.
能量既不能创造,也不能消失,尽管其形式可以转变。

2. 译成表示"无条件"的条件分句

The imitation of living systems, *be it direct or indirect*, is very useful for devising machines, hence the rapid development of bionics.
对生物的模仿不管是直接的还是间接的,对于机械设计者都很有用处,因此仿生学才迅速发展。

六、目的状语从句的译法

1. 译成表示"目的"的后置分句

英语的状语从句通常位于句末,翻译时一般采用顺译法。

A rocket must attain a speed of about five miles per second *so that it may put a satellite in orbit*.
火箭必须获得每秒大约 5mile (1mile = 1.60934km) 的速度以便把卫星送入轨道。

2. 译成表示"目的"的前置分句

汉语里表示"目的"的分句常用"为了"做关联词,置于句首,往往有强调的含意。

All the parts for this kind of machine must be made of especially strong materials *in order that they will not break while in use*.
为了使用时不致断裂,这种机器的所有部件都应该用特别坚固的材料制成。

七、结果状语从句的译法

英语和汉语都把表示"结果"的从句置于主句之后,因此这类句子可采用顺译法。注意汉译时应少用连词,或省略连词。

Electronic computers work so fast *that they can solve a very difficult problem in a few seconds*.
电子计算机工作如此迅速,一个很难的问题几秒钟内就能解决。

Fall Prevention

In the construction industry in the U. S., falls are the leading cause of worker fatalities. Each year, on average, between 150 and 200 workers are killed and more than 100,000 are injured as a result of falls at construction sites. Occupational Safety and Health Administration (OSHA) recognizes that accidents involving falls are generally complex events frequently involving a variety of factors. Consequently, the standard for fall protection deals with both the human and equipment-related issues in protecting workers from fall hazards. For example, employers and employees need to do the following:

- where protection is required, select fall protection systems appropriate for given situations;
- use proper construction and installation of safety systems;
- supervise employees properly;

- use safe work procedures;
- train workers in the proper selection, use, and maintenance of fall protection systems.

The UK Construction Industry suffered 79 fatal accidents during the year 2001/02, and 71 during 2002-2003. Almost half of these fatalities were due to "falls from heights" and around half of these were due to "falls from roof edge" and "falls through roof". The statistics also suggest that new build (e.g. profiled aluminium roofs) are just as dangerous as refurbishment (e.g. asbestos cement roofs). Falls from height are a major problem.

If safety equipment is inappropriately used, the likelihood is that it will not perform the task for which it was manufactured. It is the responsibility of the designers and planning team to identify the potential of misuse of equipment and plan for avoidance. Whether it is as simple as horseplay (e.g. jumping into nets or mats) or negligence (e.g. removing essential parts of a system), the potential dangers could be catastrophic.

When planning for the use of a fall arrest system, the planner must ensure that the decision-maker of which system will be selected has considered many factors prior to final selection, and has been furnished with enough information to make an educated decision. The following factors were considered throughout the data collection phase: the convenience of the system, installation, fittings, fixtures, comfort, traversing and travelling issues, influence of people around, etc. This text reports on interim research findings on the significance of the planning function when considering working at height.

There are five key height safety areas that a program of research is investigating. The research objectives are to evaluate the benefits and limitations of:
- fall arrest mats when working at heights and/or near a leading edge;
- safety nets during roof work;
- purlin trolley systems during industrial roof work;
- cable based fall arrest systems as a means of protection when working at heights and/or near a leading edge.

1. Fall Arrest Mats

There are situations in construction where nets are impractical and the alternative means of fall protection has normally been a harnesses and line. These include the installation of precast slabs where there is always a leading edge at each floor of the building. The concept of safety bags is growing in recognition within the industry. Passive fall protection, for example air-inflated mats, has been adopted by the Precast Flooring Federation (PFF).

Perhaps the greatest opportunity for these systems is in domestic housing. This sector of the industry has always struggled with the concept of safety nets, as the nets have their limitations when used during low level construction as they require strict management of the space below the net to ensure a clear net deflection height is maintained. Harnesses are also problematic during this type of work because it is difficult to find attachment points and workers have to attach/detach frequently. Therefore, fall arrest mats may have much to offer the housing sector. It has also become clear that a notable portion of the low-rise housing sector is opting for the soft filled mats. The popular view of problems of perception, for example, air bags being viewed as a child's amusement park toy, has been investigated and found to be generally insignificant.

2. Safety Nets

These systems are growing in popularity within the UK Construction Industry as a direct result of Regulation 6 of the Construction Regulations 1996. The systems are becoming easier to use and install as new suppliers and rigging companies enter the marketplace. Further, the systems have been widely championed. It is also possible that safety nets can be used on some refurbishment work, for example, to protect falls through fragile roof lights in circumstances where the primary protection (the solid roof light cover) is carelessly removed.

Safety nets must be installed as close as practicable under the walking/working surface on which employees are working and never more than 30 feet (9.1 m) below such levels. Defective nets shall not be used. Safety nets shall be inspected at least once a week for wear, damage, and other deterioration. The maximum size of each safety net mesh opening shall not exceed 36 square inches (230 square centimeters) nor be longer than 6 inches (15 cm) on any side, and the openings, measured center-to-center, of mesh ropes or webbing, shall not exceed 6 inches (15 cm). All mesh crossings shall be secured to prevent enlargement of the mesh opening. Each safety net or section shall have a border rope for webbing with a minimum breaking strength of 5,000 pounds (22.2 kN). Connections between safety net panels shall be as strong as integral net components and be spaced no more than 6 inches (15 cm) apart. Safety nets shall be installed with sufficient clearance underneath to prevent contact with the surface or structure below. When nets are used on bridges, the potential fall area from the walking/working surface to the net shall be unobstructed.

3. Purlin Trolley Systems

These systems have been around for a long time and are usually used in conjunction with safety harnesses. This is because, traditionally, the purlin trolley has a double handrail on the "Leading Edge" (the opposite side to that being worked on) which provides protection; but it has an open "Working Edge" (the side where the sheets will be progressively installed) and thus requires a harness attached to the trolley to prevent falls at the working edge.

A number of technological advantages have been made in this area in recent years. For instance, patented systems now manage to protect the "Working Edge" by means of a trolley which limits the open area by the provision of a horizontal barrier (attached to the trolley) which rests about six inches below the location where the roof sheet will be fixed to. This means that if someone accidentally stands on the unfixed sheet, the sheet will be caught (and therefore the worker) by the horizontal barrier. The system eliminates the need for harnesses and lanyards to be used as a safe system of roof working, and also provides an alternative to the use of nets and/or harnesses.

4. Personal Fall Arrest Systems

These consist of an anchorage, connectors, and a body belt or body harness and may include a deceleration device, lifeline, or suitable combinations. If a personal fall arrest system is used for fall protection, it must

do the following:
- limit maximum arresting force on an employee to 900 pounds (4 kN) when used with a body belt;
- limit maximum arresting force on an employee to 1,800 pounds (8 kN) when used with a body harness;
- be rigged so that an employee can neither free fall more than 6 feet (1.8m) nor contact any lower level;
- bring an employee to a complete stop and limit maximum deceleration distance an employee travels to 3.5 feet (1.07m);
- have sufficient strength to withstand twice the potential impact energy of an employee free falling a distance of 6 feet (1.8m) or the free fall distance permitted by the system, whichever is less.

The use of body belts for fall arrest is currently allowed, but effective January 1, 1998, the use of a body belt for fall arrest will be prohibited; however, the use of a body belt in a positioning device system is acceptable. Personal fall arrest systems must be inspected prior to each use for wear damage, and other deterioration. Defective components must be removed from service. Dee-rings and snap hooks must have a minimum tensile strength of 5,000 pounds (22.2 kN). Dee-rings and snap hooks shall be proof-tested to a minimum tensile load of 3,600 pounds (16 kN) without cracking, breaking, or suffering permanent deformation.

5. Planning for Safe Working at Height

Generic information throughout the data collection phase shows that planning issues must be approached in a methodical manner to ensure that equipment selection is both appropriate and effective. Specific areas considered essential in the planning process are:

In order that there are safe conditions for all operatives to carry out their function of work, the organization must accept responsibility. During construction operations, the importance of this function is crucial. The organization must take all practicable steps to ensure that all risks are designed out of the construction process at as early a stage as possible. If this is not achievable, they must utilize the expertise of those within the supply chain to fully consult and develop the submitted method statements and risk assessments to provide the workers with a safe system of work. This can involve a degree of training for all levels in the organization.

6. Safety Monitoring Systems

When no other alternative fall protection has been implemented, the employer shall implement a safety monitoring system. Employers must appoint a competent person to monitor the safety of workers and the employer shall ensure that the safety monitor is:
- competent in the recognition of fall hazards;
- capable of warning workers of fall hazard dangers and in detecting unsafe work practices;
- operating on the same walking/working surfaces of the workers and can see them;
- close enough to work operations to communicate orally with workers and has no other duties to distract from the monitoring function.

Mechanical equipment shall not be used or stored in areas where safety monitoring systems are being used to monitor employees engaged in roofing operations on low-sloped roofs. No worker, other than one engaged in roofing work (on low-sloped roofs) or one covered by a fall protection plan, shall be allowed in an area where an employee is being protected by a safety monitoring system. All workers in a controlled access zone shall be instructed to promptly comply with fall hazard warnings issued by safety monitors.

New Words and Expressions

statistics [stə'tɪstɪks]	n.	统计，统计资料
refurbishment [rɪ'fɜːbɪʃmənt]	n.	重新磨亮；整修
negligence ['neglɪdʒəns]	n.	粗心，疏忽行为
catastrophic [ˌkætə'strɒfɪk]	adj.	灾难的
purlin [pɜː'lɪn]	n.	平行桁条
alternative [ɔːl'tɜːnətɪv]	adj.	两者（或若干）中择一的
precast [priː'kɑːst]	v.	预浇铸；预制
deflection [dɪ'flekʃən]	n.	偏斜，偏向，挠曲，偏度，挠度
unobstructed ['ʌnəb'strʌktɪd]	adj.	没有障碍的，畅通无阻的
anchorage ['æŋkərɪdʒ]	n.	下锚，停泊
deceleration [diːˌselə'reɪʃən]	n.	减速
withstand [wɪð'stænd]	v.	抵挡，反抗；禁得起
low-sloped [ˌləʊ'sləʊpt]	adj.	建得很低的

Unit Thirteen

Accident Analysis in Mine Industry

Coal is produced from underground mines in about 50 countries. Underground coal mines range from modern mines using the latest remote-controlled equipment operated by a small, highly skilled workforce benefiting from continuous monitoring of all aspects of workplace conditions, to mines that are dug by hand and where the coal is extracted and transported by hand, often in conditions that are unsafe and unhealthy.

Underground coal mining has historically been one of the highest risk activities as far as the safety and health of the workforce are concerned. Fortunately, significant, sustained improvements in coal mining occupational safety and health have been achieved as a result of new technologies, massive capital investment, intensive and continuous training and changes in attitudes to safety and health among those in all stages of the coal chain. Nonetheless, if a safety net, which includes a number of critical checks and balances, is not in place to assess and control the hazards, accidents, ill health and diseases can and do occur. These are discussed as follows.

1. Rock Falls

Underground coal mines frequently suffer from roof falls which have various consequences ranging from fatalities and injuries to downtimes. Underground mining still has one of the highest fatal injury rates of any U. S. industry—more than five times the national average compared to other industries. Between 1996 and 1998, nearly half of all underground fatalities were attributed to roof, rib and face falls. Small pieces of rock falling between bolts injure 500–600 coal miners each year.

Several factors have contribution to occurrences of roof falls in underground coal mines, such as geological and stress conditions, mine layout, and mine environment. Among the factors affecting the roof fall hazards in coal mines, stress condition and mine layout are somewhat controllable by appropriate mine design. However, it is relatively more difficult to control the effect of geological conditions on roof falls, since the geological conditions are the nature's uncertainty, and hence they comprise inherent variability in roof fall occurrences. Therefore, in order to deal with the uncertainties associated with the roof falls, risk assessment methods are required for decreasing the consequences and related costs of roof fall hazards.

2. Outburst

Natural phenomena of sudden gas and rock mass outbursts, such as volcano eruptions, geysers, huge bursts of water saturated with CO_2, out of the reservoirs in former volcano craters have been known for a long time. As mining activities upset the natural balance in the rock mass containing the substances that undergo phase transitions, the outbursts of rock and gas may occur. Their occurrences in mines have been recorded for more than 150 years. Attempts have been made to provide an adequate explanation of these processes. Increasing frequency of outburst occurrence after World War II called for still more extensive research.

Coal and gas outburst problems have been exacerbated significantly over the past decades because of higher productivity and the trend towards recovery of deeper coal seams. However, despite of great efforts, surprisingly little progress has been made towards understanding outburst mechanism. Prediction techniques continue to be unreliable and unexpected outburst incidents are still a major concern for underground coal mining.

In China outbursts occur in a number of coal fields and in a large number of mines. The most important coal fields where outbursts occur are in the provinces of Shanxi (Yangquan); Liaoning (Beipiao); Henan (Jiaozuo), Chonqing (Nantong and Songzao) and Hebei (Kailuan). Coal and gas bursts are differentiated in China into four categories:

- coal bursts with no gas;
- gas bursts;
- coal and gas outbursts;
- rock and gas outbursts.

3. Mine Fires

Three ingredients are necessary for a fire. These are fuel, oxygen and ignition, referred to as the fire triangle. Coal seams make up a third of the fire triangle with natural deposits of both solid and gaseous fuels. Mine ventilation carries oxygen, the second part of the fire triangle, throughout the mine. Electrical machines, equipment, lights, power stations and circuitry, along with diesel equipment, conveyor belting frictional sources, welding, acetylene cutting and other producers of friction, spark or flame used throughout a mine are ignition sources which add the third ingredient of the fire triangle. To prevent the outbreak of coal mine fires, a number of critical safeguards, checks and balances are necessary.

Fires are a significant hazard to the safety and health of mine workers. Fires at underground and surface mines place the lives and livelihood of our nation miners at risk. Ventilation streams in underground mines can carry smoke and toxic combustion products throughout the mine, making escape through miles of confined passageways difficult and time consuming. A fire in an underground coal mine is especially hazardous due to the unlimited fuel supply and the presence of flammable methane gas. The greatest mine fire disaster in the U.S. occurred at the Cherry Coal Mine, IL, in November 1909, where 259 miners perished. During 1990-2001, more than 975 reportable fires occurred in the U.S. mining industry, causing over 470 injuries, 6 fatalities, and the temporary closing of several mines. Over 95 of the fires occurred in underground coal mines. The leading causes of mine fires include flame cutting and welding operations, friction, electrical shorts, mobile equipment malfunctions, and spontaneous combustion. The prevention, early and reliable detection, control, and suppression of mine fires are critical elements in safeguarding the lives and livelihood of over 230,000 mine workers.

4. Explosions

While much progress has been made in preventing explosion disasters in coal mines, explosions still occur, sometimes producing multiple fatalities. Explosions and the resulting fires often kill or trap workers, block avenues of escape, and rapidly generate asphyxiating gases, threatening every worker underground. Explosions in underground mines and surface processing facilities are caused by accumulations of flammable gas and/or combustible dust mixed with air in the presence of an ignition source. Explosions can be prevented by minimizing methane concentrations through methane drainage and ventilation, by adding sufficient rock dust to inert the coal dust, and by eliminating ignition sources. Explosion effects can be mitigated by using barriers to suppress propagating explosions.

5. Coal Dusts

The production, transportation and processing of coal generates small particles of coal dust. If uncontrolled and allowed to accumulate, that highly explosive dust can ignite. If it becomes airborne the coal dust can cause violent explosions. Coal dust explosions can create deadly forces, fire and super-heated air which can quickly spread through a mine, killing or injuring several miners. Explosion forces can destroy ventilation and roof controls, block escape routes and trap miners in conditions where oxygen in the mine air is replaced with asphyxiating gases.

The production, transportation and processing of coal generates tiny respirable coal dust particles that become airborne and are invisible to the naked eye. Appropriate instrumentation should be used to quantify the level and size of dust particles present in the air. Coal is made up of a variety of elements. It is mixed with other dusts, most notably crystalline silica, generated from fractured rock in the mine roof, floor or the coal seam which can also become airborne. So coal mine dusts can be a significant health risk. When inhaled by miners, dust can result in diseases of the pulmonary system (lungs), including coal-workers' pneumoconiosis (CWP), progressive massive fibrosis (PMF), silicosis, and chronic obstructive pulmonary disease (COPD). These lung diseases are progressive, disabling and can be fatal.

6. Electricity

The use of electricity and energized equipment in underground coal mines can result in injuries and death from electrical shock or arc burns. Given the confined space of underground mines, which are a dark, and at times a harsh environment, with several pieces of energized equipment and circuitry in close proximity to workers and with self-propelled equipment in motion, the potential of shock or electrocution exists.

Coal mines contain natural deposits of coal, coal mine dust and mine gases that are flammable and explosive. The introduction of electrical and energized equipment in coal mines creates the potential of igniting mine fires and explosions, which can cause numerous deaths and injuries from single events and devastate the mine.

Electrical accidents are the 4th leading cause of death in mining and are disproportionately fatal compared with most other types of mining accidents. About one-fifth of these deaths result when high-reaching mobile equipment contacts power lines overhead. One-half of all mine electrical injuries and fatalities occur during electrical maintenance work, with the following electrical components most commonly involved: circuit breakers, conductors, batteries, and meters.

7. Inrushes of Water, Gas or Other Material

Inrushes of water, noxious or flammable gas or other materials are a serious hazard in coal mining. Mining operations can get too close to old workings or geological abnormalities that contain water, gases or materials that could inundate the mine. One particular hazard is mining next to old workings that were poorly surveyed, not surveyed at all, or not adequately inspected, which contain bodies of water or dangerous mine gases. Old workings filled with water, particularly at elevations higher than the active mine, could quickly flood the mine and drown miners before they could escape if inadvertently cut into. Inrushing mine gases inadvertently encountered can overpower mine ventilation and the oxygen in the air and suffocate miners or, with the right mixture of oxygen, trigger explosions.

New Words and Expressions

extract [ɪkˈstrækt]	v. 开采
rock fall	岩石冒落
downtime [ˈdaʊntaɪm]	n. 停工
rib [rɪb]	n. 矿柱
face [feɪs]	n. 采煤工作面
bolt [bəʊlt]	n. 锚杆
stress [stres]	n. 应力
layout [ˈleɪˌaʊt]	n. 方案，布局
outburst [ˈaʊtbɜːst]	n. 突出
geyser [ˈgaɪzə]	n. 间歇泉
crater [ˈkreɪtə]	n. 火山口
phase transition	相变
exacerbate [eksˈæsɜːbeɪt]	v. 加剧
mechanism [ˈmekənɪzəm]	n. 机理
coal field	煤田
mine fire	矿井火灾
acetylene [əˈsetɪliːn]	n. 乙炔气
flammable [ˈflæməbl]	adj. 易燃的
methane [ˈmiːθeɪn]	n. 甲烷
spontaneous combustion	自然发火
asphyxiate [æsˈfɪksɪeɪt]	v. 使……窒息

concentration [ˌkɒnsen'treɪʃən]	n.	浓度
drainage ['dreɪnɪdʒ]	n.	抽放
inert [ɪ'nɜːt]	adj.	惰性的
barrier ['bærɪə]	n.	隔爆物
crystalline silica		结晶二氧化硅
pulmonary ['pʌmənəri]	adj.	肺部的
pneumoconiosis ['njuːmə,kəʊnɪ'əʊsɪs]	n.	尘肺病
fibrosis [faɪ'brəʊsɪs]	n.	纤维症，纤维化
silicosis [ˌsɪlɪ'kəʊsɪs]	n.	硅肺病
chronic ['krɒnɪk]	adj.	慢性的；延续很长的
shock [ʃɒk]	v.	电击
arc burns		电弧灼伤
inrush ['ɪnrʌʃ]	n.	涌入
noxious ['nɒkʃəs]	adj.	有害的
old working		老采空区，老窑
abnormality [ˌæbnɔː'mælətɪ]	n.	畸形，异常性
inundate ['ɪnəndeɪt]	v.	淹没
survey [sɜː'veɪ]	v.	勘查
elevation [ˌelɪ'veɪʃən]	n.	开采水平（指高程）
suffocate ['sʌfəkeɪt]	v.	使……窒息

Translation Skill

科技英语翻译技巧（十二）——长句的译法

长句常见的翻译方法有：化整为零，分译法；纲举目张，变序法；逆流而上，逆序法。此外，某些长句也可能根据原文的顺序，保持不变，一气呵成，依次译出，即递序而下，顺序法。现分别举例说明。

一、化整为零，分译法

原句包含多层意思，而汉语习惯一个小句表达一层意思。为了使行文简洁，将整个长句译成几个独立的句子，顺序基本不变，保持前后的连贯。

Steel is usually made where the iron ore is smelted, so that the modern steelworks forms a complete unity, taking in raw materials and producing all types of cast iron and steel, both for sending to other works for further treatment, and as finished products such as joists and other consumers.

[初译] 通常在炼铁的地方就炼钢，所以现代炼钢厂从运进原材料到生产供送往其他工厂进一步加工处理并制成如工字钢及其他商品钢材的成品而形成一整套的联合企业。

上述这种译文读起来佶屈聱牙，看起来概念不清。究其原因，囿于英语结构形式，忽略汉语自身规律。试将原文拆散为三个独立的小句译成汉语。steel is...smelted 为第一小句；so that...steel 为第二小句；both for...consumers 为第三小句。原文中，通过 both...and 连接的两个介词短语在译文中可扩展成句子。

[改译] 通常炼铁的地方也炼钢。因此，现代炼钢厂是一个配套的整体，从运进原料到生产各种类型的铸铁与钢材；有的送往其他工厂进一步加工处理，有的就制成成品，如工字钢及其他一些成材。

The loads a structure is subjected to are divided into dead loads, which include the weights of all the parts of the structure, and live loads, which are due to the weights of people, movable equipment, etc.

[初译] 一个结构物受到的荷载可分为包括结构物各部分重量的静载和由于人及可移动设备等的重量引起的活载。

从理解原文的角度看，译文传达了原意，符合"信"的原则。但由于汉语习惯用小句，这46个字的长句很难一气读完，有欠"达、雅"。如采用化整为零，分译法，则可达醒目易读之效。

[改译] 一个结构物受到的荷载可分为静载与活载两类。静载包括该结构物各部分的重量。活载则是由于人及可移动设备等的重量而引起的荷载。

二、纲举目张，变序法

原句结构复杂，可按汉语由远及近的顺序从中间断句，层层展开，最后画龙点睛，突出主题。

An "alloy" steel is one which, in addition to the contents of carbon, sulphur and phosphorus, contains more than 1% of manganese, or more than 0.3% of silicon, or some other elements in amounts not encountered in carbon steels.

[初译] 合金钢是一种钢。除掉碳、硫、磷以外，还含有多于1%的锰或多于0.3%的硅或者一些碳素钢中不包括的其他元素。

以上译文采用了分译法。但"合金钢是一种钢"这句话毫无价值，没有再现作者的原意。而且与后半部分脱节，失去逻辑上的严密性。试从 in addition to... 译起，最后回到句首，展示主要信息，既可衔接紧密，又能突出主题。

[改译] 如果一种钢除含有碳、硫、磷以外，还含有多于1%的锰或多于0.3%的硅或者一些碳素钢中不包含的其他元素，那么这种钢便是"合金钢"。

The reason that a neutral body is attracted by a charged body is that, although the neutral body is neutral within itself, it is not neutral with respect to the charged body, and the two bodies act as if oppositely charged when brought near each other.

[初译] 中性物体被带电物体吸引的原因在于，虽然中性物体本身是不带电的，但对带电体来说，它不是中性。当这两个物体彼此接近时，就会产生带有相反电荷的作用。

上述译文不恰当地采用了顺序分译法，以致译文内部衔接松弛，破坏了作者所提出的概念的完整性。从 although... 到句尾，都是说明"中性物体被带电体吸引的原因"，这一整体不容分割。试从 although 引导的让步状语从句入手，将原文前置的主要信息在译文中后置，画龙点睛。

[改译] 虽然中性物体本身是不带电的，但对于带电体来说，它并非中性；当这两个物体彼此接近时，就会产生极性相反的电荷的作用。这就是中性物体被带电体吸引的原因。

三、逆流而上，逆序法

由于英语惯用前置性陈述，先果后因；而汉语相反，一般先因后果，层层递进，最后综合，点出主题。处理这类句子，宜于先译全句的后部，再依次向前，逆序译出。

The construction of such a satellite is now believed to be quite realizable, its realization being supported with all the achievements of contemporary science, which have brought into being not only materials capable of withstanding severe stresses involved and high temperatures developed, but new technological processes as well.

[初译] 制造这样的人造卫星确信是可能的，因为可以依靠现代科学的一切成果。这些成果不

仅提供了能够承受高温高压的材料,而且也提供了新的工艺过程。

原文由三部分构成:主句,做原因状语的分词独立结构,修饰独立结构的定语从句。根据汉语词序,原因状语在先,定语前置,故从 which... 入手,再译出 its realization... 最后才译出 The construction... realizable。

[改译] 现代科学的一切成就不仅提供了能够承受高温高压的材料,而且也提供了新的工艺过程。依靠现代科学的这些成果,我们相信完全可以制造出这样的人造卫星。

In reality, the lines of division between sciences are becoming blurred, and science is again approaching the "unity" that it had two centuries ago—although the accumulated knowledge is enormously greater now, and no one person can hope to comprehend more than a fraction of it.

[初译] 事实上,各学科之间的分界线变得模糊不清,科学再次近似于两百年前那样的"单一整体"——虽然现在积累起来的知识比以往多得多,而且任何个人也只可望了解其中的一小部分。

翻译一是要传达原旨,二是要符合汉语习惯。上述译文虽然达旨,但是西化汉语。应遵照汉语的习惯,将由 although 引出的让步状语从句提前,逆序而上为好。

[改译] 虽然现在积累起来的知识要多得多,而且任何个人也只可望了解其中的一小部分,但事实上,各学科之间界线却变得模糊不清,科学再次近似于两百年前那样的"单一整体"。

四、递序而下,顺序法

英语原句的结构的顺序与汉语相同,层次分明,译成汉语时可顺序推进,一气呵成。

All commercial iron and steel contains iron as chief constituent, but the percentages of carbon and other elements and the methods by which iron and steel is produced, as well as the processes to which they may be subjected, so change the characteristic properties that there are many distinct forms of iron and steel, some of which have properties so different as to appear like different metals.

[译文] 所有商用钢铁都以其含铁为主要成分,但由于炭和其他元素的含量不同,钢铁冶炼方法不同以及加工过程不同,从而改变了它们的特性,以至于有多种不同的钢铁,其中有些钢铁的特性极不相同,看上去就像不同的金属一样。

The development of rockets has made possible the achievement of speeds of several thousand miles per hour, and what is more important it has brought within reach of these rockets heights far beyond those which can be reached by aeroplanes, and where there is little or no air resistance, and so it is much easier both to obtain and to maintain such speed.

[译文] 火箭技术的进展已使速度可达每小时几千英里,而更为重要的是,这种进展已使火箭所能达到的高度大大超过了飞机所能达到的高度,在这样的高度上,几乎没有空气阻力,因而很容易达到并保持火箭的那种高速度。

Reading Material

Explosions in Gobs in Coal Mines

In the last few years, seven explosions of methane and/or coal dust occurred within worked-out, sealed areas (gobs) of underground U.S. coal mines. These explosions, believed to have been started by lightning, destroyed many mine seals and caused much damage occurred in one mine in Alabama over a 3-year period. Fortunately, the explosion forces and the toxic gases that vented from the sealed gob area did not cause fatalities or injuries, but they destroyed several large seals. If miners had been inspecting or working near these areas, the potential for serious injury and/or death would have been high. The United Mine Workers of America and the Mine Safety and Health Administration (MSHA) requested that the National Institute for Occupational Safety and Health (NIOSH) help identify the mechanisms for lightning penetration into the gob and recommend ways to reduce the probability of occurrence of future explosions from such lightning penetrations into sealed areas of underground coal mines. Because three of the explosions occurred in a single mine, special attention was focused on these particular explosions. The first occurred in April 1994 in a sealed area, which enclosed about 1.35 square miles of abandoned workings (gob). This explosion destroyed 3 of the 38 seals that surrounded the gob. These seals were less than 2,000 ft. (1 ft = 0.3048 m) away from three 4.5-in-diam steel-cased test wells that extended from the surface into the mine entry. At the time of the explosion, the National Lightning Detection Network (NLDN) verified 12 lightning strikes within 10 miles of the mine, including several above the gob. After the 1994 explosion, the damaged seals were rebuilt using an acceptable seal design capable of withstanding a 20-psi explosion, as required by 30 CFR 75.335. On January 26, 1996, a second gob explosion destroyed five more cementitious pumpable seals less than 2,000 ft. away from those destroyed in 1994 and even closer to the steel-cased wells. The NLDN verified 72 lightning strikes in the area of this second gob explosion. Compressive strength analyses of fragments from the destroyed seals showed that strengths ranged from 11 to 138 psi, with an average of 83 psi. This is over 100 psi below the minimum 200-psi compressive strength requirement for cementitious pumpable seals. These seals were again replaced. On July 9, 1997, the third and most violent explosion occurred in the same vicinity of the last two explosions. Three more cementitious pumpable seals were destroyed, including one newer seal that exceeded the minimum 200-psi compressive strength requirement. The NLDN verified 695 lightning strikes above the mine during the time in which the explosion occurred. MSHA's accident investigation report indicated that the maximum pressure (propagating forces) of this explosion exceeded 20 psi. The current seal construction regulations, which relate to all suitable underground seal designs, assume that an explosion occurring in the gob will not be stronger than 20-psi pressure. However, if a large flammable gas volume exists in the gob, the resulting explosion pressure can be more than 20 psi.

　　Two conditions are necessary for an explosion to occur: ①a fuel-air mixture with a fuel concentration in the flammable (explosible) range; ②an effective source of ignition for that mixture. If the source of ignition, in this case presumed to be the lightning, cannot be eliminated, then the only alternative is to eliminate flammable concentrations or reduce the volume of the flammable mixture present in the sealed area. If the gob atmosphere contains methane concentrations greatly above the upper flammable limit of 15% methane, it will be unaffected by lightning or by other potential sources that might exist in the gob. Other ignition sources could include the sudden discharge of old batteries, roof falls, and spontaneous combustion. After the third explosion in the same gob area in the Alabama mine, all parties agreed to pressure balance the sealed gob area in order to reduce the leakage of air into the gob and therefore to increase the concentration of methane in the gob to well above the upper flammable limit. By reducing the average differential pressure from 3 in (1 in = 0.0254 m). to about 0.46 in. of water gauge across the gob, the average volumetric air leakage into the sealed area was reduced by a factor of 2.5. This reduction in seal leakage greatly reduced the probability of formation of a large volume of flammable methane-air mixture in the gob.

The sealed gob area was pressure balanced in July 1997, and the methane concentrations in the gob were monitored by MSHA through February 2000. The gas samples from behind the seals indicated methane levels more than 20%. This is well above the upper flammable limit. During this same period, several severe storms, accompanied by lightning, passed over the gob area without apparently triggering a gob explosion. One storm was so severe that an imbedded tornado passed over the mine, ripping the doors from the ventilation fan without disturbing the gob. To date, it seems that the combination of constructing suitable seals that met the requirements of the CFR, coupled with pressure balancing of the sealed gob, helped reduce gas leakage and minimized the formation of a flammable methane-air mixture in the gob.

Minimizing pressure differentials across seals should be considered an essential part of the overall strategy for sealing gobs. Reducing the pressure differential reduces the air leakage through the seal and thus reduces the formation of large flammable methane-air volumes in the gob. Any wires or metal conductors, including steelcased wells, that connect the surface and the gob area should be removed. These contribute to the transfer of energy into the mine. If possible, nonconductive well casings should be used. In addition, old batteries, which are another potential ignition source, should not be left behind in the gob. During the sealing process, adequate rock dust (80% incombustible) should be used both inside and outside of the sealed areas to reduce the contribution of coal dust to a methane explosion.

New Words and Expressions

sealed [siːld]　　　　　　　　　adj. 密封的
gob [gɒb]　　　　　　　　　　　n. 凝块,(黏质物)一块
strike [straɪk]　　　　　　　　　n. 打击,攻击
cementitious [siːmənˈtɪʃəs]　　adj. 似水泥的;有黏性的
ignition [ɪgˈnɪʃən]　　　　　　　n. 点火,点燃
battery [ˈbætəri]　　　　　　　　n. 重创,连续打击
triggering [ˈtrɪgə]　　　　　　　n. 起动;触发;控制

Unit Fourteen

Hazardous Chemical and Its Identication

1. Hazardous Chemical

The Occupational Safety and Health Administration (OSHA) defines a hazardous chemical as any chemical which is a physical hazard or a health hazard. Physical hazard means a chemical for which there is evidence that it is a combustible liquid, a compressed gas, explosive, flammable, an organic peroxide, an oxidizer, pyrophoric, unstable (reactive), or water-reactive. Health hazard means a chemical for which there is evidence that acute (immediate) or chronic (delayed) health effects may occur in over-exposed people. Exposure being related to the dose (how much), the duration and frequency of exposure (how long and how often), and the route of exposure (how and where the material gets in or on the body), whether it be absorption through: the respiratory tract (inhalation); the skin; the digestive tract (ingestion), and/or percutaneous injection through the skin (e. g. accidental needle stick). These health effects can be: transient, persistent, or cumulative, local (at the sight of initial contact with the substance) and/or systemic (after absorption, distribution, and possible biotransformation, at a site distant from initial contact with the substance). The term "health hazard" includes chemicals which are carcinogens, toxic or highly toxic, reproductive toxins, irritants, corrosives, sensitizers, hepatotoxins, nephrotoxins, neurotoxins, agents which act on the hematopoietic system, and agents which damage the lungs, skin, eyes, or mucous membranes.

Although safety hazards related to the physical characteristics of a chemical can be objectively defined in terms of testing requirements (e. g. flammability), health hazards definitions are less precise and more subjective. Physical hazards may manifest as fires, explosions, excessive temperatures, or the release of large volumes of gas or toxic or flammable gases or vapors. Health hazards, depending on the exposure, may cause measurable changes in the body, such as decreased pulmonary (lung) function. These changes are generally indicated by the occurrence of signs and symptoms in the over-exposed person, such as shortness of breath, a non-measurable, subjective feeling.

There have been many attempts to categorize health effects and to define them in various ways. Generally, the terms "acute" and "chronic" are used to delineate between effects on the basis of severity or duration.

"Acute" effects usually occur rapidly as a result of short-term exposures, and may be of short duration. "Chronic" effects generally occur as a result of long-term exposure, and may be of long duration.

2. Identification of Hazardous Chemical

There have been several instances of major industrial disasters related to the use of chemicals. Although they are individual accidents, different in the way in which they happened and the chemicals that were involved, they have one common feature: They were uncontrolled, involving fires, explosions or the release of toxic substances that either resulted in the death and injury of large numbers of people inside and outside of the factory or caused extensive damage to the property and the environment. Actually, accidents involving major hazards could start with:

- leakage of a flammable substance, mixing of the substance with air, formation of a flammable vapour cloud and drifting of the cloud to a source of ignition leading to a fire or an explosion;
- leakage of toxic substances, formation of a toxic vapour cloud and drifting of the cloud.

These clouds would directly affect the site as well as possibly the surrounding populated areas. In the case of flammable substances the greatest danger arises from sudden massive escape of volatile liquids or gases. If the cloud were ignited, the effects of combustion would depend on many factors, such as wind speed and the extent to which the cloud was diluted. The area affected would generally be limited to a few hundred meters from the site.

Much larger areas can be dangerously affected in a sudden release or by very large quantities of toxic materials. In favorable conditions such a cloud can still contain lethal concentrations of toxic chemicals several kilometers from the accident site. The extent of casualty depends on the number of people in the path of the cloud and on the efficiency of emergency arrangements, for example, evacuation before the cloud reaches the populated areas.

The effect can also migrate into other factories situated nearby and containing flammable, reactive or toxic chemicals, escalating the disaster. This is sometimes referred to as the "domino effect".

Not only does the cloud itself pose a health hazard, but the fires cause depletion of oxygen and fumes generated by the fire may contain toxic gases. Chlorine and ammonia are the toxic chemicals most commonly used in quantities large enough to pose a major hazard. Both have a history of major accidents. There are also other chemicals which, although used in smaller quantities should, be handled with particular care because of their higher toxicity.

An industrial accident classified as a "major hazard" leads to tighter control, more specific than that applied in the normal factory operations. This is in order to protect both workers and outside people, to avoid economical losses to the factory and damage to the environment.

The first step in a systematic approach is to identify the installations susceptible to a "major hazard". For this purpose, EU in Europe has a directive which has been in use since 1984. The directive sets certain criteria based on the toxic, flammable and explosive properties of the chemicals. For the selection of specific industrial activities which involve a major hazard, a list of substances with limit amounts is provided. The list contains 180 toxic substances with the limits varying from 1 kg for extremely toxic substances to 50,000 tons for highly flammable liquids.

3. Criteria for Major Hazard Substances

(1) **Very Toxic and Toxic Substances**

Substances classified to hazard categories below according to their acute toxicity. Classification can also be done by determining the acute toxicity in animals, expressed in LD50 or in LC50 values and using the following limits:

- Substances which correspond to the first line of the table below.
- Substances which correspond to the second and third line and which, owing to their physical and chemical properties, are potential candidates for a major hazard similar to that caused by substances filling the criteria of the first line in the table.

Category	LD50 absorbed orally in rat (mg/kg bodyweight)	LD50 dermal absorption in rat or rabbit (mg/kg bodyweight)	LC50 absorbed by inhalation in rat (mg/litre per 4 hours)
1	<5	<10	<0.10
2	5~25	10~50	0.1~0.5
3	25~200	50~400	0.5~2

(2) **Flammable Substances**

- Gases which form flammable mixtures with air.
- Highly or extremely flammable liquids with flash points lower than 21 F.
- Flammable liquids with flash points lower than 55 F.

Substances which may explode when in contact with a source of ignition or which are more sensitive to shock and friction than dinitrobenzene. The industrial activities creating the risk of a major hazard may not be restricted to defined sectors. Experience has shown that such installations are most commonly associated with the following activities and workplace:

- petrochemical works in refineries;
- chemical works in chemical production plants;
- lPG (Liquid Petroleum Gas) storage and terminals;
- stores and distribution centres of chemicals;
- factories handling explosives;
- works in which chlorine is used in bulk quantities.

New Words and Expressions

peroxide [pəˈrɒksaɪd]　　　　　　　　n. 过氧化物
pyrophoric [ˌpaɪərəʊˈfɒrɪk]　　　　　adj. 发生火花的，生火花的
respiratory [rɪsˈpaɪərətəri]　　　　　adj. 呼吸的，呼吸系统的
respiratory tract　　　　　　　　　　呼吸道
inhalation [ˌɪnhəˈleɪʃən]　　　　　　n. 吸入
percutaneous [ˌpɜːkjuːˈteɪnɪəs]　　　adj. 经由皮肤的
cumulative [ˈkjuːmjʊlətɪv]　　　　　adj. 累积的，累积性的

carcinogen [kɑː'sɪnədʒən]	n.	致癌物（质），诱癌因素
irritant ['ɪrɪtənt]	n.	刺激性物质，刺激性气体
hematopoietic [hemə'tɒpɪətɪk]	adj.	造血的，生血的
mucous ['mjuːkəs]	adj.	黏液的，黏液似的
pulmonary ['pʌlmənəri]	adj.	肺部的
delineate [dɪ'lɪnɪeɪt]	v.	叙述，描写
volatile ['vɒlətaɪl]	adj.	挥发（性）的
dilute [daɪ'ljuːt]	v.	稀释，冲淡
lethal ['liːθəl]	adj.	致命的，致死的
casualty ['kæʒjʊəlti]	n.	事故，灾难；死伤
evacuation [ɪ,vækjʊ'eɪʃən]	n.	撤退，撤散，疏散
migrate [maɪ'greɪt]	v.	迁移
escalate ['eskəleɪt]	v.	使逐步升级（加剧）
depletion [dɪ'pliːʃən]	n.	消耗，损耗
fume ['fjuːm]	n.	(pl.) 烟气（雾），浓烟
dinitrobenzene [daɪ,naɪtrəʊ'benziːn]	n.	二硝基苯

Translation Skill

科技应用文的译法

说明书、合同、协议等科技应用文有其独特的文体及相应的格式，翻译这类应用文首先要熟悉它们的常用句型及表达方式，才能正确理解，翻译通顺。

一、说明书常用词语、句型及其译法

1. 常用词语及其译法

operational instructions, operating instructions, service instructions, user's instructions, operation manual, instruction manual, working manual, operating manual 使用说明书；general 概述；construction 构造；features 特点；design 结构；group designation 总类名称；transportation 搬运；main assembles and controls 主要部件及操纵机构；coolant system 冷却系统；inspection 检验；electric system 电气系统；location 安装地点；description 主要部件；coarse adjustment 粗调；installation 安装；first commissioning 试车；adjustment 调试；instructions for erection 安装规程；fine adjustment 微调节器；service condition 工作条件；test run 试运转；directions for use 用法；major operating components and their functions 主要操作部件及其功能；system diagram 系统示意图；wiring/circuit diagram 线路图；rated capacity 额定容量；rated load 额定负载；operating flow chart 操作流程图；power requirements 电源条件；maintenance 维修；operating voltage 工作电压；list of stuffing and wearing parts 填充料及易磨损零件表；factory services 工厂检修；data book 数据表；wear adjustment 易损部分的调整；working drawings of ease-worn parts 易损零件图；cleaning 清洗；high voltage cautions 小心高压电；precautions/cautions 注意事项；dimensions 尺寸；specifications 规格；length 长；width 宽；depth 深；height 高；weight 重；net weight 净重；gross weight 毛重；measurement 尺码；nominal speed 额定转速；measuring range 量程；factor 安全系数；accessories/accessories supplied 附件；work cycle 工作周期；accessory case 附件箱；oiling period 加油时间；operating humidity 工作湿度；standard accessories 标准附件；operating temperature 工作温度；code of accessories 附件代号。

2. 常用句型及其译法

Place the new pump in position and tighten mounting bolts.
使水泵就位，拧紧装配螺柱。

Make sure hoses are clamped tightly.
务必将皮管夹紧。

The Voltage Selector Setting should be checked to see that it conforms to the local AC supply voltage.
必须查看本机交流电压选择之预调状态是否符合本地交流电压。

Be careful to install batteries in series.
要注意，所有电池应装成串联。

Do not allow any buttons to remain depressed when the unit is not in use, as this may result in damage to the pinch roller.
本机不使用时，任何控制键钮不得留在锁定状态，以免使压轮受到损伤。

Never use cleaning fluids, chemicals or wax.
切勿使用洗涤剂、化学药品及石蜡。

What step to take in these cases...
如何处理下述故障……

Adjust the volume of the MIC, monitor speaker/phone and room speaker until they are at a proper level.
将麦克风、监听扬声器/耳机以及室内扬声器的音量调整至适当电平。

If the record player is equipped with an earth lead, connect it to the GND terminal of the unit.
该电唱机如备有接地线，则应将它与本机的接地端子连接起来。

Remove the AC supply lead before servicing or cleaning heads, rollers, etc.
切断交流电源才能维修、清洗磁头、压轮等部件。

For headphones listening, connect the headphones plug into the "phones" socket.
如用头戴耳机收听，请在标有"phone"字样的插座连接头戴耳机。

Specification are subject to change without notice.
规格如有变更，恕不另行通知。

To obtain the best performance and ensure years of trouble-free use, please read this instruction manual carefully.
请仔细阅读说明书，以便使本机发挥其最佳性能，经久耐用，不出故障。

二、合同常用词语、句型及其译法

1. 常用词语及其译法

contract 合同；confirmation 确认书；contract of employment 聘约合同；contract of trade 贸易合同；sales confirmation 成交确认书；contract of purchase 定购合同；name of commodity 商品名称；specification 规格；quantity 数量；unit price 单价；total value 总值；packing 包装；country of origin and manufacturer 生产国别及制造厂商；terms of payment 付款条件；insurance 保险；time of shipment 装运时间；port of loading 装运口岸；port of destination 目的口岸；shipping marks 装运标记；other terms 其他条款；the buyers 买方；the sellers 卖方；the engaging party 聘方；hereafter to be called the first party (the second party) 以下简称甲方（乙方）; the engaged party 受聘方; the term of service 聘期; expiration of the contract 合同到期; to renew the contract 延长合同期; guarantee period 保险期; extension of employment contract 延长聘期合同; transshipment 转船; partial shipment 分装; compensation allowance 补偿津贴。

2. 常用句型及其译法

The undersigned sellers and buyers have agreed to close the following transactions according to the terms and conditions stipulated below.

兹经买卖双方同意成交下列商品，特签订条款如下。

To be covered by the buyers.

由买方负责。

To be effected by the sellers covering all risks and war risk for 1.5% of invoice value.

由卖方按发票总值的 1.5% 抽保综合险及战争险。

By irrevocable letter of credit available against opening bank's receipt of the documents in compliance with the credit terms and conditions as stipulated.

开不可撤销信用证，开证银行收到符合信用证条款及本合同规定之单据后付款。

On each package shall be stenciled conspicuously: port of destination, package number, gross and net weights, measurement and the shipping mark.

每件货物应明显地标出口岸、件号、毛重、净重、尺码及装运标记。

Any claim shall be lodged within 60 days from the date of import.

自进口日期起索赔权 60 天。

The present contract is made out in Chinese and English, both versions being equally valid.

本合同由中文和英文两种文字写成。两种文本具有同等效力。

Neither party shall cancel the contract without sufficient cause of reason.

双方均不得无故解除合同。

Should any other clause in this contract be in conflict with the following supplementary conditions, the supplementary conditions should be taken as final and binding.

本合同之其他任何条款如与本附加条款有抵触时，以本附加条款为准。

三、协议常用词语、句型及其译法

1. 常用词语及其译法

agreement on academic exchange 学术交流协议; agreement on technological co-operation 技术合作协议; agreement on production co-operation 生产合作协议; agreement on personnel training 人员培训协议; agreement on the import of a colour kinescope production line 引进彩色显像管生产线协议书; co-operative agreement on science and technology 科技合作协议; patterns and contents of co-operation 合作方式与内容; measures of implementation 执行措施; personnel matters 人员交流; financial arrangements 费用安排; duration of agreement 协议有效期限; signature 签字; the two sides 双方; on the basis of mutual benefit 在互惠基础上; to establish closer co-operation in technological transfers and information exchange 在技术转让与互通信息方面建立更为密切的合作关系; joint venture 合资; Sino-foreign joint venture 中外合资; foreign exchange 外汇; foreign currencies 外币; technical consultation 技术咨询; technical training 技术培训

2. 常用句型及其译法

... hereby indicate their intention to enter into a program of technological co-operation to benefit both companies. ……特此表明双方在互惠基础上签订技术合作协议的愿望。

to provide equipment, engineering technology, technicians and managerial personnel including quality inspectors 提供设备、工程技术、技术人员以及包括质量检验员在内的管理人员

to conclude an agreement as follows 特签订以下协议

to go into effect from the date of signature 自签字之日起生效

to be effective immediately when signed 自签字之日起立即生效

... must be completed 18 months after the conclusion of the present agreement……限在本协议签订后一年半内建成

the take-over will take place... 某月某日验收

This agreement is hereby made on the basis of the existing contract and consultation of both sides. 双方根据已有的联系与协商，特签订如下协议。

The two sides agree to co-operate with each other in research projects of common interest. 双方同意对共同感兴趣的项目进行合作。

Detail provision concerning such co-operation will be worked out later through consultation. 有关合作细则，由双方另行商定。

The provisions of this agreement may be amended at any time upon written consent of the participating co-operators. 合作双方在任何时候可通过书面协商，同意对本协议进行修改。

At is expiration, the agreement may be modified or the period of validity may be extended through mutual consultation. 协议期满后，通过协商，可予修改或延长有效期。

The agreement is to be executed in English and Chinese versions. Both texts are equally valid. 本协议有英文、中文两种文本，两种文本具有同等效力。

Reading Material

Basic Principles for Controlling Chemical Hazards

Chemicals are considered highly hazardous for many reasons. They may cause cancer, birth defects, induce genetic damage, cause miscarriage, or otherwise interfere with the reproductive process. Or they may be a cholinesterase inhibitor, a cyanide, or other highly toxic chemical that, after a comparatively small exposure, can lead to serious injury or even death. Working with compounds like these generally necessitates implementation of additional safety precautions.

The goal of defining precisely, in measurable terms, every possible health effect that may occur in the workplace as a result of chemical exposures cannot realistically be accomplished. This does not negate the need for laboratory personnel to know about the possible effects as well as the physical hazards of the hazardous chemicals they use, and to protect themselves from these effects and hazards. Controlling possible hazards may require the application of engineering hazard controls (substitution, minimization, isolation, ventilation) supplemented by administrative hazard controls (planning, information and training, written policies and procedures, safe work practices, and environmental and medical surveillance). Personal protective equipment (e.g. gloves, goggles, coats, respirators) may need to be considered if engineering and administrative controls are not technically, operationally, or financially feasible. Typically, combinations of all three will be necessary to control the hazards.

It should also be kept in mind that the risks associated with the possession and use of a hazardous chemical are dependent upon a multitude of factors, all of which must be considered before acquiring and using a hazardous chemical. Important elements to examine and address include: the knowledge of and commitment to safe laboratory practices of all who handle the chemical; its physical, chemical, and biological properties and those of its derivatives; the quantity received and the manner in which it is stored and distributed; the manner in which it is used; the manner of disposal of the substance and its derivatives; the length of time it is on the premises, and the number of persons who work in the area and have open access to the substance (the Preliminary Chemical Hazard Assessment Form can be used as part of this risk assessment). The decision to procure a specific quantity of a specific hazardous chemical is a commitment to handle it responsibly from receipt to ultimate disposal.

1. Basic Hazard Control Rationale

The basic principles for controlling chemical hazards can be broken down into three broad categories: engineering controls, administrative controls, and personal protective equipment. Hazards must be controlled first by the application of engineering controls that are supplemented by administrative controls. Personal protective equipment is only considered when other controls are not technically, operationally or financially feasible. Typically, combinations of all methods are necessary in controlling chemical hazards.

2. Engineering Hazard Controls

Engineering hazard controls may be defined as an installation of equipment, or other physical facilities including, if necessary, the selection and arrangement of experimental equipment. Engineering controls remove the hazard, either by initial design specifications or by applying methods of substitution, minimization, isolation, or ventilation. Engineering controls are the most effective hazard control methods, especially when introduced at the conceptual stage of planning when control measures can be integrated more readily into the design. They tend to be more effective than other hazard controls (administrative controls and personal protective equipment) because they remove the source of the hazard or reduce the hazard rather than lessen the damage that may result from the hazard. They are also less dependent on the chemical user who, unfortunately, is subjected to all of the frailties which befall humans (e.g. forgetfulness, preoccupation, insufficient knowledge).

Substitution refers to the replacement of a hazardous material or process with one that is less hazardous (e.g. the replacement of mercury thermometers with alcohol thermometers or dip coating materials rather than spray coating to reduce the inhalation hazard). Minimization is the expression used when a hazard is lessened by scaling down the hazardous process. Hence, the quantity of hazardous materials used and stored is reduced, lessening the potential hazards. Isolation is the term applied when a barrier is interposed between a material, equipment or process hazard and the property or persons who might be affected by the hazard.

Substitution is usually the least expensive and the most positive method of controlling hazards and should always be the first engineering hazard control measure considered. Minimization should always be the next engineering control measure attempted after examining substitution followed by the consideration of isolation. Isolation is particularly useful when the material, equipment or process requires minimal contact or manipulations. When these previously mentioned control methods are not feasible, ventilation is the next desirable engineering option.

Ventilation is used to control toxic and/or flammable atmospheres by exhausting or supplying air to either remove hazardous atmospheres at their source or dilute them to a safe level. The two types of ventilation are typically termed local exhaust and general ventilation. Local exhaust attempts to enclose the material, equipment or process as much as possible and to withdraw air from the physical enclosure at a rate sufficient to assure that the direction of air movement at all openings is always into the enclosure. General ventilation attempts to control hazardous atmospheres by diluting the atmosphere to a safe level by either exhausting or supplying air to the general area.

Local exhaust is always the preferable ventilation method but is typically more costly. For some situations, general ventilation may suffice but only if the following criteria are met: only small quantities of air contaminants are released into the area at fairly uniform rates; there is sufficient distance between the person and the contaminant source to allow sufficient air movement to dilute the contaminant to a safe level; only materials of low toxicity or flammability are being used; there is no need to collect or filter the contaminant before the exhaust air is discharged into the environment (including the rest of the building), and the contaminant will not produce corrosion or other damage to equipment in the area or in any way affect other building occupants outside the general use area.

3. Administrative Hazard Controls

All of the aforementioned engineering hazard control methods, in order to exist or be effective, require the application of "administrative hazard controls" as either supplemental hazard controls or to ensure that engineering controls are developed, maintained, and properly functioning. Administrative hazard controls consist of managerial efforts to reduce hazards through planning, information and training (e.g. hazard communication), written policies and procedures (e.g. the Chemical Hygiene Plan), safe work practices, and environmental and medical surveillance (e.g. work place inspections, equipment preventive maintenance, and exposure monitoring). Because they primarily address the human element of hazard controls, they are of vital importance and are always used to control chemical hazards.

4. Personal Protective Equipment

As was mentioned earlier, when adequate engineering and administrative hazard controls are not technically, operationally, or financially feasible, personal protective equipment must be considered as a supplement. Personal protective equipment (PPE) includes a wide variety of items worn by an individual to isolate the person from chemical hazards. PPE includes articles to protect the eyes, skin, and the respiratory tract. In some situations, PPE may be the only reasonable hazard control option, but for many reasons it is the least desirable means of controlling chemical hazards. PPE users must be aware of, and compensate for these undesirable qualities. PPE does not eliminate hazards but merely minimizes damage from hazards. The effectiveness of PPE is highly dependent on the user. PPE is oftentimes cumbersome and uncomfortable to wear. Each type of PPE has specific applications, advantages, limitations, and potential problems associated with their misuse and those using PPE must be fully knowledgeable of these considerations. PPE must match the hazards and the conditions of use and be properly maintained in order to be effective. Their misuse may directly or indirectly contribute to the hazard or create a new one. The material of construction must be compatible with the chemical's hazards and must maximize protection, dexterity, and comfort.

5. Every Hazard Can Be Controlled

Not all the previously mentioned principles are applicable to controlling the hazards of every chemical, but all chemical hazards can be controlled by the application of at least one of these methods. Ingenuity, experience, and a complete understanding of the circumstances surrounding the control problem will be required in choosing methods which will not only provide adequate hazard control, but which will consider development, installation, and/or operating costs as well as human factors such as user acceptance, convenience, comfort, etc.

New Words and Expressions

cholinesterase [ˌkɒlɪˈnestəreɪs]	n.	胆碱酯酶
cyanide [ˈsaɪənaɪd]	n.	氰化物
necessitate [nɪˈsesɪteɪt]	v.	使成为必需
surveillance [sɜːˈveɪləns]	n.	监测，监督
commitment [kəˈmɪtmənt]	n.	承诺，应允的义务
derivative [dɪˈrɪvətɪv]	n.	衍生物
disposal [dɪsˈpəʊzəl]	n.	处置，处理，处理方式
facility [fəˈsɪlɪti]	n.	设施，设备
frailty [ˈfreɪlti]	n.	意志薄弱，性格缺陷
befall [bɪˈfɔːl]	v.	落到……的身上，降临于
preoccupation [priːˌɒkjʊˈpeɪʃən]	n.	偏见，成见
mercury thermometer		水银温度计
scale down		降低，减小
interpose [ˌɪntəˈpəʊz]	v.	放入，插入
manipulation [məˌnɪpjʊˈleɪʃən]	n.	处理，操作
enclose [ɪnˈkləʊz]	v.	密封，围包住
contaminant [kənˈtæmɪnənt]	n.	污染物，致污物
corrosion [kəˈrəʊʒən]	n.	腐蚀，侵蚀
cumbersome [ˈkʌmbəsəm]	adj.	麻烦的；笨重的
dexterity [deksˈterɪti]	n.	（手）灵巧，熟练
ingenuity [ˌɪndʒɪˈnjuːɪti]	n.	机灵；独创性；精巧

Unit Fifteen

Combustion and Explosion Accidents

Flammable substances present a substantial hazard in the form of fires and explosions. The combustion of one gallon of toluene can destroy an ordinary chemistry laboratory in minutes; persons present may be killed. The potential consequences of fires and explosions in plants and plant environments are even greater. The direct losses resulting from fires and explosions are substantial. Additional losses in life and business interruptions are also substantial.

1. Combustion and Explosion

Burning is the rapid exothermic oxidation of an ignited fuel. The essential elements for combustion are fuel, an oxidizer, and an ignition source. These elements are illustrated by the fire triangle. When fuel, oxidizer, and an ignition source are present at the necessary levels, burning will occur. Two common examples of the three components of the fire triangle are wood, air, and a match; and methane, air, and a spark. However, other, less obvious combinations of chemicals can lead to fires and explosions.

The fuel can be in solid, liquid, or vapor form, but vapor and liquid fuels are generally easier to ignite. The combustion always occurs in the vapor phase; liquids are volatized and solids are decomposed into vapor before combustion. It is worth noting that many solid materials including common metals such as iron and aluminum become flammable when reduced to a fine powder. The rapid combustion of fine solid particles may lead to dust explosions.

Various oxidizers can also be found in fire and explosion accidents especially in chemical industry, such as oxygen (or air), fluorine, chlorine and hydrogen peroxide, nitric acid, perchloric acid, etc. Sparks, flames, static electricity, heat are the most common ignition sources. Ignition of a flammable mixture may be caused by a flammable mixture coming in contact with a source of ignition with sufficient energy or the gas reaching a temperature high enough to cause the gas to auto-ignite.

Flammable gas or vapor and air mixtures will ignite and burn only over a well-specified range of compositions. The mixture will not burn when the composition is lower than the lower flammable limit (LFL); the mixture is too lean for combustion. The mixture is also not combustible when the composition is too rich; that is, when it is above the upper flammable limit (UFL). A mixture is flammable only when the composition is between the LFL and the UFL. Lower explosion limit LEL and upper explosion limit UEL are used interchangeably with LFL and UFL.

The flash point of a liquid is the lowest temperature at which it gives off enough vapor to form an ignitable mixture with air. At the flash point the vapor will burn but only briefly; inadequate vapor is produced to maintain combustion. The flash point generally increases with increasing pressure. There are several different experimental methods used to determine flash points. The two most commonly used methods are open cup and closed cup, depending on the physical configuration of the experimental equipment. The open-cup flash point is a few degrees higher than the closed-cup flash point. The fire point is the lowest temperature at which a vapor above a liquid will continue to burn once ignited; the fire point temperature is higher than the flash point.

2. Accidental Explosion

The damage effects from an explosion depend highly on whether the explosion results from a detonation or a deflagration. The difference depends on whether the reaction front propagates above or below the speed of sound in the unreacted gases. In some combustion reactions the reaction front is propagated by a strong pressure wave, which compresses the unreacted mixture in front of the reaction front above its autoignition temperature. This compression occurs rapidly, resulting in an abrupt pressure change or shock in front of the reaction front. This is classified as a detonation, resulting in a reaction front and leading shock wave that propagates into the unreacted mixture at or above the sonic velocity. For a deflagration the energy from the reaction is transferred to the unreacted mixture by heat conduction and molecular diffusion. These processes are relatively slow, causing the reaction front to propagate at a speed less than the sonic velocity. A deflagration can also evolve into a detonation. This is called a deflagration to detonation transition (DDT). The transition is particularly common in pipes but unlikely in vessels or open spaces. In a piping system energy from a deflagration can feed forward to the pressure wave, resulting in an increase in the adiabatic pressure rise. The pressure builds and results in a full detonation.

Confined explosion is an explosion occurring within a vessel or a building. These are most common and usually result in injury to the building inhabitants and extensive damage. Unconfined explosions occur in the open. This type of explosion is usually the result of a flammable gas spill. The gas is dispersed and mixed with air until it comes in contact with an ignition source. Unconfined explosions are rarer than confined explosions because the explosive material is frequently diluted below the LFL by wind dispersion. These explosions are destructive because large quantities of gas and large areas are frequently involved.

The most dangerous and destructive explosions in the process industries are vapor cloud explosions (VCEs). These explosions occur in a sequence of steps:
- Sudden release of a large quantity of flammable vapor (typically this occurs when a vessel, containing a superheated liquid and pressurized liquid, ruptures).
- Dispersion of the vapor throughout the plant site while mixing with air.
- Ignition of the resulting vapor cloud.

The accident at Flixborough, England, is a classic example of a VCE. A sudden failure of a 20 in cyclohexane line between reactors led to vaporization of an estimated 30 tons of cyclohexane. The vapor cloud dispersed throughout the plant site and was ignited by an unknown source 45 seconds after the release. The entire plant site was leveled and 28 people were killed.

VCEs have increased in number because of an increase in inventories of flammable materials in process plants and because of operations at more severe conditions. Any process containing quantities of liquefied gases, volatile superheated liquid, or high-pressure gases is considered a good candidate for a VCE.

Some of the parameters that affect VCE behaviors are quantity of material released, fraction of material vaporized, probability of ignition of the cloud, distance traveled by the cloud before ignition, time delay before ignition of cloud, probability of explosion rather than fire, existence of a threshold quantity of material, efficiency of explosion, and location of ignition source with respect to release. From a safety standpoint the best approach is to prevent the release of material. A large cloud of combustible material is dangerous and almost impossible to control, despite any safety systems installed to prevent ignition. Methods that are used to prevent VCEs include keeping low inventories of volatile, flammable materials, using process conditions that minimize flashing if a vessel or pipeline is ruptured, using analyzers to detect leaks at low concentrations, and installing automated block valves to shut systems down while the spill is in the incipient stage of development.

The major distinction between fires and explosions is the rate of energy release. Fires release energy slowly, whereas explosions release energy rapidly, typically on the order of microseconds. Fires can also result from explosions, and explosions can result from fires. For example, a Boiling-Liquid-Expanding-Vapor Explosion (BLEVE) occurs when a tank containing a liquid held above its atmospheric pressure boiling point ruptures, resulting in the explosive vaporization of a large fraction of the tank contents. The most common type of BLEVE is caused by fire. The energy released by the BLEVE process itself can result in considerable damage. If the materials are flammable, a VCE might result.

3. Explosion Effects

An explosion results from the rapid release of energy. The energy release must be sudden enough to cause a local accumulation of energy at the site of the explosion. This energy is then dissipated by a variety of mechanisms, including formation of a pressure wave, projectiles, thermal radiation, and acoustic energy. The damage from an explosion is caused by the dissipating energy.

The explosion of gas or dust cloud results in a reaction front moving outward from the ignition source preceded by a shock wave or pressure front. After the combustible material is consumed, the reaction front terminates, but the pressure wave continues its outward movement. A blast wave is composed of the pressure wave and subsequent wind. It is the blast wave that causes most of the damage.

An explosion occurring in a confined vessel or structure can rupture the vessel or structure, resulting in the projection of debris over a wide area. This debris, or missiles, can cause appreciable injury to people and damage to structures and process equipment. Unconfined explosions also create missiles by blast wave impact and subsequent translation of structures. Missiles are frequently a means by which an accident propagates throughout a plant facility. A localized explosion in one part of the plant projects debris throughout the plant. This debris strikes storage tanks, process equipment, and pipelines, resulting in secondary fires or explosions. People can be injured by explosions from direct blast effects (including overpressure and thermal radiation) or indirect blast effects (mostly missile damage).

Unit Fifteen Combustion and Explosion Accidents

New Words and Expressions

combustion and explosion	燃烧与爆炸
fluorine ['fluəri:n]	n. 氟
hydrogen peroxide	过氧化氢
nitric acid ['naɪtrɪk 'æsɪd]	n. 硝酸
perchloric acid [pə'klɔ:rɪk 'æsɪd]	n. 高氯酸
ignition source	点火源
auto-ignite	自动着火
lower flammable limit (LFL)	可燃下限
upper flammable limit (UFL)	可燃上限
flash point	闪点
reaction front	反应阵面
confined explosion	受限（或约束）爆炸
unconfined explosion	非受限（或非约束）爆炸
vapor cloud explosions (VCEs)	气云爆炸
superheated liquid	过热液体
pressurized liquid	加压液化气体
cyclohexane [,saɪklə(ʊ)'heksein]	n. 环己烷
Boiling-Liquid-Expanding-Vapor Explosion (BLEVE)	沸腾液体膨胀汽化爆炸
shock wave	冲击波，激波
blast effect	爆炸波效应

科技英语摘要的写作要点

摘要是对整篇研究论文的简要性概述，以精炼的文字对论文的主要内容进行概括，是一篇浓缩的研究短文。摘要的文字应简洁，句式要完整，前后句式要连贯，是整篇文章的精华。摘要的主要内容一般包括：研究的目的、研究的方法、获得的结果或成果、结论及研究结果的价值。

一、语法要点

1）在叙述研究目的时，句子既可使用一般现在时态，也可使用一般过去时态。

- The purpose of this paper is to obtain a new model of safety evaluation for mechanical industry.
- This paper focuses on the challenges facing for small enterprises in relation to occupational safety and health.
- The purpose of this article is to explore the measurement for weak magnetic fields present in the human body.
- The purpose of this study was to identify macro-socioeconomic determinants for the incidence of occupational injuries and diseases among home care workers.
- The objective of this research was to optimize the performance of device at high temperature.

2）摘要的内容是介绍分析方法或技术、数学模型、算法时，研究方法的叙述应用一般现在时态。
- The finite element method is used to simulate the velocity of gas flow in mine.
- The cavity length control is done by moving the prism along its symmetry axis using piezoelectric element.

如果叙述的内容是试验程序或试验方法，通常使用一般过去时态。
- A suspension of carbonyl iron particles in silicone oil and mineral oil with the volume fraction from 10% to 35% respectively was used in experiment.

3）使用一般过去时态叙述主要研究结果；但作者认为自己的研究结果普遍有效，而不仅仅是在本次研究的特定条件下才适用，则可以使用一般现在时态。
- It is suggested that some basic steps be taken to control incident rates in occupational safety and health management.
- The evaluation system of organization safety culture characterizes all the features as much as safety culture in an enterprise.
- The paper concludes that the new method for selecting the sampling period is desirable.

二、摘要常用句型

This paper presents a thorough study of the input/output stability theory of relay pulse senders.
本文深入研究了中继脉冲发送器的输入-输出稳定性的理论。

This paper treats an important problem in database management systems.
本文讨论了数据管库理系统的一个重要问题。

This paper addresses problems in linear quadratic optimal control of EHT transmission line.
本文论述了超压输电线路线性二次优化控制的若干问题。

This paper establishes convergence properties of a new algorithm.
本文论证了一种新算法的收敛性。

This paper shows results on multiple time-scale system.
本文展示了对多时标系统研究的成果。

This paper is concerned with the derivation of optimum data...
本文介绍了……最佳数据的方法。

This paper discusses the relation between the sampling period and the stability of sampled-data.
本文讨论了抽样周期与抽样数据稳定性之间的相互关系。

This paper considers the design of controllers for flexible systems.
本文研究了通用系统控制器的设计。

This paper describes a new algorithmic language.
本文介绍了一种新的算法语言。

This paper proposes (develops, extends) a new approach for the analysis of tolerance.
本文提出了一种分析容错的新方法。

This paper provides a new framework for the analyses of loss of power.
本文提出了一个功率损耗分析方法的框架。

The purpose of this article is to explore the measurement for weak magnetic fields present in the human body.
本文旨在研究测量人体内弱磁场的方法。

The author describes a new techniques of noise control.

作者介绍了一种噪声控制新技术。

In this paper, the parameter imbedding algorithm is introduced.

本文介绍了一种参数插入算法。

The paper concludes that the new method for selecting the sampling period is desirable.

本文的结论是：这种选择采样周期的新方法是可取的。

Reading Material

Prevention and Protection for Dust Explosion

1. Dust Explosion

While fire is caused when three factors—fuel, oxidant, and ignition—come together to make what has been called "the fire triangle", a dust explosion demands two more factors: mixing (of dust and air), and confinement (of the dust cloud). The "dust explosion pentagon" is formed when these five factors occur together: ①presence of combustible dust in a finely divided form; ②availability of oxidant; ③presence of an ignition source; ④some degree of confinement; ⑤state of mixed reactants. A point to be noted here is that even partial confinement of an ignited dust cloud is sufficient to cause a highly damaging explosion. In this sense, too, dust clouds behave in a manner similar to clouds of flammable gases.

In case of dusts made up of volatile substances, the explosion may occur in three steps which may follow each other in very quick succession—devolatization (where volatiles are let off by the particle or the particles are vapourized), gas phase mixing of fuel (released by dusts) and oxidant (usually air), and gas phase combustion. Many combustible dusts if dispersed as a cloud in air and ignited, will allow a flame to propagate through the cloud in a manner similar to (though not identical to) the propagation of flames in premixed fuel-oxidant gases. Such dusts include common foodstuffs like sugar flour, cocoa, synthetic materials such as plastics, chemicals and pharmaceuticals, metals such as aluminum and magnesium, and traditional fuels such as coal and wood. Generally dust explosion involves oxide formation. But metal dusts can also react with nitrogen or carbon dioxide to generate heat for explosion.

Explosion hazard always exists whenever dusts are produced, stored or processed and where situations can occur when these materials are present as a mixture in air. The mixture is deemed "explosible" if combustible dusts are present in such quantities in air that an explosion can occur on ignition. More than 70% of dusts processed in industry are combustible. This implies that majority of industrial plants that have dust-processing equipment are susceptible to dust explosions. A dust explosion is initiated by the rapid combustion of flammable particulates suspended in air. Any solid material that can burn in air will do so with a violence and speed that increases with the degree of sub-division of the material. Higher the degree of sub-division (or smaller the particle size) more rapid and explosive the burning, till a limiting stage is reached when particles too fine in size tend to lump together. If the ignited dust cloud is unconfined, it would only cause a flash fire. But if the ignited dust cloud is confined, even partially, the heat of combustion may result in rapid development of pressure, with flame propagation across the dust cloud and the evolution of large quantities of heat and reaction products. Besides the particle size, the violence of such an explosion depends on the rate of energy release due to combustion relative to the degree of confinement and heat losses. In exceptional situations a destructive explosion can occur even in an unconfined dust cloud if the reactions caused by combustion are so fast that pressure builds up in the dust cloud faster than it can be dissipated at the edge of the cloud. The oxygen required for combustion is mostly supplied by air. The condition necessary for a dust explosion is a simultaneous presence of dust cloud of appropriate concentration in air that will support combustion throughout the process and a suitable ignition source.

Even though mention of dust explosions is found in literature since 1785, systematic records are available only from the early 20th century. One of the earliest recorded and the most serious of the accidents triggered by dust explosion occurred at Leiden, the Netherlands, on January 12, 1807. The explosion killed 151 and wounded about 2000. Houses collapsed up to a distance of 155 m from the explosion source and within the whole city people were hit by flying debris, glass, and roof tiles. The catastrophic dust explosion which occurred at the Harbin Linen Textile Plant, People's Republic of China on March 15, 1987, killed 58 persons and injured another 177. It destroyed 13,000 m^2 of factory area. The ignition was possibly caused by an electrostatic spark in one of the dust collecting units. The explosion then propagated through the other seven dust collecting units, demolishing most of the plant. An explosion and fire involving polyethylene dust killed 6 workers and injured 38 others at the USA-based West Pharmaceuticals (Kinston, NC) on January 29, 2003. Two firefighters were among those killed in a massive blast of which impact was felt over a large area; the burning debris triggered secondary fires up to 2 miles away. Some initiating event caused dust to become airborne above a suspended ceiling. There it contacted an ignition source leading to the catastrophic event.

Indeed it is commonly admitted that at least one dust explosion occurs in each industrialized country every day, and dust explosions has been recognized as very major industrial hazards that can match or exceed the ferocity of well-known industrial disasters like the one that occurred in Flixborough, and meticulous attention is paid towards their analysis, prevention, and control.

2. Dust Explosion Prevention Strategies

The right conditions that must prevail for a dust explosion to occur is summed up under the "dust explosion pentagon". The most obvious way to prevent a dust explosion from happening is to not allow the dust pentagon to be closed. This can be attempted in the following ways:
- effectively modifying the process to reduce dust handling hazards;
- preventing suspensions of flammable dusts;
- completely removing or minimizing the presence of ignition sources;
- inerting.

(1) **Process Modification**

The most obvious way to prevent dust explosions is to replace existing processes with the ones which do not deal with combustible dusts, that is so called inherent safe method. Inherently safe process design to prevent or reduce dust explosion hazard involve use of such production, treatment, transportation and storage operations where dust cloud generation is kept at a minimum. One example is use of mass flow silos and hoppers instead of the frequently used funnel flow types.

(2) **Preventing Flammable Dust Suspensions**

It is difficult to keep the flammable dust cloud concentrations below certain levels in order to prevent an explosion, because the minimum explosive concentration is usually far below the economic operational conditions. The following measures may be effective: ①In cases where high dust concentration may be unavoidable, it would be appropriate to work with smaller piles of dust than with one large one. ②Situations such as the free fall of dust from a height into a hoper, which may encourage dust cloud formation, should be avoided. ③The dust removal process, say from a gas stream, must be done at as early a stage as process considerations permit in order to avoid dust suspensions. ④Plants handling flammable dusts should be appropriately designed to minimize the accumulation of dusts. Cleaning of dusts collected in places like ducts should be facilitated as often as permissible. It must be emphasized that even if a dust suspension within the explosive range is not present during normal operations, it may be so during startup, shutdown or fault conditions. Once a dust explosion is initiated, the expanding gases behind the flame of such incipient dust explosion can whirlup the otherwise settled dust lying nearby, thus feeding the explosion. By adhering to certain safe housekeeping practices, the presence of dust can be limited to controlled locations thereby reducing the potential for the formation of hazardous dust clouds.

(3) **Elimination of Ignition Sources**

In situations where the minimum electrical spark ignition energy of the working dust is considerably greater than 10 MJ, elimination of ignition sources would provide adequate protection against dust explosions. The ignition sources, which are traceable to routine operations or worker habits such as smoking, open flames, open light (bulbs), welding, cutting, and grinding, can be eliminated by sufficient staff training and enforcement of discipline. The ignition sources that originate in the process itself involve factors such as open flames, hot surfaces, self-heating, smouldering nests and exothermic decomposition, heat from mechanical impacts, exothermic decomposition of dust via mechanical impacts, and electric sparks and electrostatic discharges. As these ignition conditions are inherent in the actual process, the hazard can be reduced by employing the right precautionary measures like earthing of equipment that may develop charges and strict adherence to the process operation norms.

(4) **Inerting**

"Inerting" refers to ways and means by which the oxygen concentration in a process area or a vessel is reduced by adding an inert gas to a level at which the dust cloud can no longer propagate a self-sustaining flame. Such inerting would slow down or totally prevent the dust explosion pentagon from taking shape, thereby reducing the explosion hazard. "Inerting" is also practiced, though much less frequently, by mixing a combustible dust with a non-combustible one. The gases commonly used for inerting of hazardous dusts are nitrogen, carbon dioxide, water vapour and rare gases. Selecting a suitable gas depends on various factors, the principle one being the reactivity of a gas with the dust for which it is used.

The system is slightly evacuated and then flushed with the inert gas until the original pressure is regained. This is repeated until the desired level of inerting is accomplished. If a high pressure system is being used, the inert gas may simply be pumped into the process vessels until the desired pressure is reached. Once inerting has been done, care must be taken that no air leaks into the process. If a new gas is introduced with the feed, it should also be inerted. Often partial inerting is used where total inerting may be too costly; this does not eliminate the chance of explosion, but limits it substantively. To accomplish partial inerting the gas (most often air) in which the explosible dust is dispersed is mixed with a fraction of inert gas (e.g. nitrogen) considerably smaller than that required for complete inerting. This reduces both the explosibility and the ignition sensitivity of the dust cloud.

3. Dust Explosion Damage Control

Due to the myriad and complex ways in which dust explosions can occur, it is more or less impossible to eliminate the dust explosion hazard. But control measures can drastically reduce the damage caused by the explosions, both in terms of lesser property losses, and lesser process shutdown time. The dust explosion damage control strategies revolve round:

- explosion containment;
- explosion isolation;
- explosion suppression;
- explosion venting.

(1) **Explosion Containment**

If a dust explosion can be contained within a designated space, much of the damage it may cause to the surroundings can be controlled. Containment is an attractive option, since it is an essentially passive method and avoids the problem of relief disposal. It is not usually practicable, however, to design the whole of a dust handling plant so that it can withstand the pressures generated by dust explosions. This is particularly the case with large plants. Nevertheless containment is practicable in small-scale units and on certain equipment. A grinding mill, for example, can be made strong enough to withstand a dust explosion. The maximum explosion pressure for most flammable gases and dusts is in the range 7–10 barg. But the static pressure is not the sole criterion; the rate of pressure rise in a dust explosion being high, the equipment must be able to withstand this dynamic loading. The equipment should be designed on the basis of rotational symmetry and avoid large flat surfaces and angular parts. Particular attention should be paid to the points at which dust is fed or withdrawn from the plant and to the connections between units. When the powder/dust is highly toxic, complete and reliable confinement is absolutely necessary.

(2) **Explosion Isolation**

The objective of explosion isolation is to prevent dust explosions from spreading from the primary explosion location to other process units, workrooms, etc. Due to pressure-piling, jet-initiated high initial turbulence and turbulent jet ignition, very high pressure peaks can be generated even in generously vented vessels necessitating effective means of explosion isolation in inter-connected systems. Two approaches are commonly adopted: use of quick acting shut-off valves, and material chokes. For explosion isolation involving quick acting shut-off valves, the valves are installed in pipes connecting one vessel with another. The valves are activated by explosion detectors which are equipped with pressure and/or optical sensors. The former type is usually preferred, since an optical detector can be blinded. But a pressure sensor may not detect very weak pressure waves which an optical sensor is capable of detecting. The time required for the valve to close depends on the distance between the remote pressure or flame sensor and the valve.

(3) **Explosion Suppression**

If, in process equipment which harbor dust explosion hazard, a system can be put in place which gets activated as soon as an explosion begins to occur, suppresses it by swiftly adding suitable inertants, and prevents it from re-building, the risk of explosions can be greatly reduced. The mechanisms of suppression of the explosion are: quenching, free radical scavenging, wetting and inerting. Of these the principal mechanisms are quenching, or abstraction of heat. Automatic explosion suppression devices aim to achieve this objective. An explosion suppression system must have four basic attributes: ①It should respond to an explosion with minimum time delay by getting activated quickly. ②It should inject a suppressant in adequate quantities within a very short time in a manner as to counter the incipient explosion and arrest the propagating flame. ③To shut down the plant. ④To prevent the plant from getting restarted until the explosion hazard has been mitigated. The common inertants include Halons, water and dry chemical suppressants. The design of a suppression system being a complex function of the triggering pressure, geometry of the area to be protected, nature of suppressant, the suppression system hardware, etc. Simple design nomograms and equations, based on the "cubic law" of the maximum rate of pressure rise, have also been derived to aid the design engineer in assessing the effectiveness of explosion suppression in practice.

(4) Explosion Venting

When all attempts to prevent a dust explosion have failed, the explosion would occur. If the explosion can be vented effectively, its adverse impact can be minimized. Except when toxic dusts are involved, venting can significantly reduce the destructive potential of a dust explosion. By sizing such a vent properly, it may be ensured that the vent becomes operative as soon as the overpressure exceeds a certain safe threshold and sufficient quantities of gas (and particulates) are let off quickly to prevent the pressure in the protected area from reaching destructive levels. To size the vent area one needs an understanding of all the factors which determine the severity of a dust explosion, including the geometry of the unit in which the provision of explosion venting is being made, dust concentration, initial pressure, initial temperature, initial turbulence, ignition source, presence of flammable gas or inert gas/dust. All these factors influence the explosion pressure—in terms of rate as well as extent of pressure rise. In addition one needs to consider the reduced explosion pressure, vent opening pressure, vent area, vent distribution, vent opening, and vent panel.

Furthermore, protecting the vented enclosure is not the only important concern in dust explosion venting. The blast waves and flames which are emitted into the surroundings by the vent can be hazardous. The maximum flame length emitted from a vent can be up to 10 times the cube root of the vented vessel volume. If too much unburned flammable material is ejected by a vent it may even get ignited by the vented flame to cause a secondary explosion. The challenge of vent design thus extends to eliminating hazardous effects of vented material, especially the flames.

New Words and Expressions

fire triangle	火灾三角形
dust explosion	粉尘爆炸
confinement [kənˈfaɪnmənt]	n. 限制，约束
explosion hazard	爆炸危险性
explosible [ɪkˈspləʊzəbl]	adj. 容易爆炸的，可爆炸的
particulate [pɑːˈtɪkjʊlət]	n. 微粒，粒子，颗粒物
flash fire	闪燃火灾
firefighter [ˈfaɪəfaɪtə]	n. 消防队员

secondary fire	二次火灾
ferocity [fəˈrɒsɪti]	n. 暴行，凶猛，残暴
meticulous [məˈtɪkjələs]	adj. 谨小慎微的，过度重视细节的
silo [ˈsaɪləʊ]	n. 筒仓，储塔
hopper [ˈhɒpə]	n. 送料斗，加料斗
minimum explosive concentration	最低爆炸浓度
explosive range	爆炸浓度范围
exothermic decomposition	放热分解
electrostatic discharge	静电放电
inert [ɪˈnɜːt]	adj. 惰性的，不活泼的
self-sustaining flame	自持传播火焰
explosion containment	抗爆，承爆
explosion isolation	爆炸隔绝
explosion suppression	爆炸抑制
grinding mill	研磨机，磨碎机，破碎机
pressure-piling	压力叠加
turbulent jet ignition	流射流点火
quick acting shut-off valves	快速动作截止阀
quench [kwentʃ]	v. （用水）扑灭（火焰等），熄灭，猝熄

Unit Sixteen

The History of Nuclear Power Plant Safety

Safety has been an important consideration from the very beginning of the development of nuclear reactors. On December 2, 1942, when the first atomic reactor was brought to criticality, Enrico Fermi had already made safety an important part of the experiment. In addition to a shutoff rod, other emergency procedures for shutting down the pile were prepared in advance. Fermi also considered the safety aspects of reactor operation. Shortly before the reactor was expected to reach criticality, Fermi noted the mounting tension of the crew. To make sure that the operation was carried out in a calm and considered manner, he directed that the experiment be shut down and that all adjourn for lunch. With such leadership in safety at the very beginning, it is no wonder that the operation of reactors to date has such an impressive track record.

When did the concept of "Nuclear Power Safety" first arise? While the world would have to do what until 1957 for the first commercial nuclear power plant, the ground work for nuclear safety began with the first major investigation into a controlled nuclear fission chain reaction. That was performed by Enrico Fermi at the University of Chicago in 1942.

Key to unleashing this new source of energy was emphasis on the "controlled" part. Just like the early New World explorers, these pioneer scientists had to prepare for the unknown. The experiment, shown at right, was an array of uranium-oxide rods embedded in graphite blocks. There plan initiated the "defense-in-depth" philosophy to safety; however this terminology was not established at the time. This concept conceded the possibility of a single failure.

To address the possibility of a failure, multiple safeguard were designed into the experiment. In the "pile" were three sets of control rods. The primary set was not used for safety at all, it was designed for fine control of the nuclear chain reaction. The other two control rods served the safety functions. One set was automatic and could be controlled by manual interaction and the other was an emergency safety rod. The automatic control rod was operated by an electric motor and responded to a "high" instrument reading from a radiation counter. Attached to one end of the emergency rod was a rope running through the pile and weighted heavily on the opposite end. During testing, this rod was withdrawn from the pile and tied down by another rope. It was the job of the "Safety Control Rod Axe Man" to stand-by ready to cut this rope with an axe should something unexpected happen, or in case the automatic safety rods failed. The acronym SCRAM from "Safety Control Rod Axe Man" is still used today in reference to the rapid shutdown of a nuclear reaction.

The safety measures did not stop with the control rods. Not wanting to rely completely on mechanical devices, Fermi organized a "liquid-control squad". They were to stand on a platform above the pile and respond to mechanical failure of the control rods by pouring a Cadmium-salt solution over the experiment. Cadmium is a strong absorber of the neutrons required to fission the uranium.

Fortunately, Fermi and their team had done their homework. The experiment went off without any problems and at 3:25pm on December 2, 1942, the nuclear age was born. The first man-made self-sustaining nuclear reaction had been achieved.

Following the success of that day, nuclear power research focused on the development of the atomic bomb as part of the "Manhattan Project". This work was performed primarily at the Los Alamos National Laboratory in New Mexico. While there is a broad difference between nuclear weapons and nuclear power plants, a great deal of nuclear theory was generated for the war effort. We know from Edward Teller that this work was managed in a safe manner. He says, "Fortunately, 'a nuclear accident' did not happen. Car accidents occurred on the road winding up the mesa; people were injured while riding horses. But the atomic materials were handled with great care and with complete safety."

Are the application of nuclear power peaceful? To many people the awesome destruction of the atomic bombs over Japan suggested more than just a new modern-age weapon of mass destruction. Nuclear physicist Alvin Weinberg told the Senate's Special Committee on Atomic Energy in December 1945: "Atomic power can cure as well as kill. It can fertilize and enrich a region as well as devastate it. It can widen man's horizons as well as force him back into the cave." Despite these calls from prominent technical leaders such as Weinberg, the U.S. government maintained strict control over atomic technology and focused research on military purposes. *The Atomic Energy Act of 1946* established the five-member Atomic Energy Commission (AEC) in 1946 and acknowledged in passing the potential peaceful benefits of atomic power. The 1946 law did not permit the privatization of nuclear power applications; rather, the government would hold a monopoly of the technology.

Today, with the Department of Energy's release of formerly classified information, it is generally believed that researchers sponsored by AEC at times breached the boundaries of ethics in regard to "studying the atom". Much of this work was performed to "understand" weapons effects and potential medical responses and had very little to do with nuclear power or the industry that came latter. This behavior continues to haunt the nuclear power industry today. In this regard, the government has forsaken the trust of it's constituents and it will take a long time before the American constituents will forgive it's government. Meanwhile, the industry developed under that veil of secrecy will suffer from the sins of it creators.

Despite the news stories of renegade scientists breaching ethical standards in the name of science, some safety issues were on the minds of the government that was supporting this new discover. Two of the biggest issues that remain with the nuclear industry today, siting and containment, were issues from the start. In the earliest large reactors, the plutonium production reactors at Hanford, the role of geographic isolation in protecting the safety of the general public was emphasized. At its first meeting in 1947, the Reactor Safeguards Committee of the AEC considered the first proposal for a contained reactor, the SIR, which was to be enclosed in a large spherical shell at West Milton, New York. From that time on, containment for protection of the general public has played an important role in reactor safety in the United States.

During the late fifties and early sixties, nuclear power went from a novel and developing technology to a commercially competitive energy source. With the rapid rise in demand for nuclear power, the AEC was inundated with a flood of licensing applications. Given that the normal licensing review time took nearly a year to complete, a backlog grew and further delays followed. One of the big issues that surfaced was reactor siting. Some of the licensing applications made proposals to site reactors rather close to metropolitan areas. Until 1960, the most projects where developed in remote locations; however, this was not always a practical option for commercial nuclear power.

Beginning in 1961, the AEC began defining a standard regulatory prescription to licensing. The reactor siting issue was the first subject addressed with the new approach. In what would become 10 CFR 100, the AEC began debate on a baseline for what would be required elements in a reactor site. ACRS member Dr. Clifford Beck began this debate with the following assumptions:

- the probability of a major accident was relatively small;
- an upper limit of fission product release could be estimated;
- reactors were expected to be in inhabited areas;
- the containment building would hold.

From these basic assumptions, Dr. Beck outlined a reactor siting criteria that focused on quantitative limits. This leads to the beginnings of a draft criteria. In this draft criteria was the first "sample calculation" for determining dose to a population. The calculation included assumptions for the limiting accident, the Source Term (amount of radioactive substances in the core), the Dispersal of the Radioactivity (including weather considerations), and an evaluation of radiation effects on people.

The solution to this type of calculation provided parameters for determining the exclusion area (no population) and a low-population zone. Within the low population zone, consequences of the maximum credible accident would be limited to dose limits of 25 rem whole body and 300 rem to the thyroid.

The reactor siting discussion brought recognition to a change in the perceived maximum credible accident. For reactors approved for research, the maximum credible accident would likely follow the careless addition of reactivity to the core resulting in a reactor excursion that would result in a intense pulse of radiation that would threaten reactor staff and followed by release of radionuclides to the atmosphere, threatening local population. Commercial designs maintained engineered limiting controls on the rate of reactivity that could be introduced; hence, the criticality excursion was not really credible for commercial plants. Instead, the loss-of-coolant-accident (LOCA)—likely the result of a major pipe break—began to dominate AEC and Advisory Committee on Reactor Safeguards (ACRS) meetings. Early calculation should that even following an SCRAM, a large break in the reactor coolant system could leave the reactor core fuel vulnerable to failure and melting. The loss of fuel integrity would release radionuclides out the break and the loss of coolant out the break could threaten the containment integrity.

Prior to the end of the 1960s, the AEC viewed the containment building as the final independent line of defense against the release of radiation. It was generally accepted that even during a severe accident, the consequences would only be felt within the containment. In 1967, a special task force commissioned by the AEC to look into the problem of core melting presented their findings that showed that under certain severe accident conditions the integrity of the containment could be breached. This finding forever changed the U. S. government's approach to nuclear power plant regulations. Regulatory focus began to shift from containment design to preventing accidents severe enough to threaten containment. This was the job of a properly designed and functioning Emergency Core Cooling Systems (ECCS).

In early 1971, some tests in the Semi-Scale Facility at the Idaho National Reactor Testing Station suggested that design of Westinghouse's PWR ECCS was deficient. The experiments were run by heating a simulated, nine-inch core electrically, allowing the cooling water to escape, and then injecting the emergency coolant. To the surprise of the investigators, the high steam pressure that was created in the vessel by the loss of coolant blocked the flow of water from the ECCS. Without even reaching the core, about 90 percent of the emergency coolant bypassed the core and went out the simulated pipe break.

To address the safety issue the Commission decided to publish the "interim acceptance criteria" for emergency cooling systems that licensees would have to meet. It imposed a series of requirements that it believed would ensure that the ECCS in a plant would prevent a core melt after a loss of coolant accident. The AEC did not prescribe methods of meeting the interim criteria, but in effect, it mandated that manufacturers and utilities set an upper limit on the amount of heat generated by reactors. In some cases, this would force utilities to reduce the power rating of their plants.

When the implications of the Semiscale experiments reached the public, complaints about the AEC's handling of the issue ranged from polite criticism to calls for a licensing moratorium and a shutdown of the eleven plants then operating. Activists from the Union of Concerned Scientists (UCS) sharply criticized the adequacy of the interim criteria. Scientists at the AEC's national laboratories, without endorsing the alarmist language that the UCS used, shared some of the same reservations. This influenced the Atomic Safety and Licensing Board to announced that in light of the uncertainties about ECCS and the interim criteria, it lacked sufficient information to approve new license applications. This opened the way for intervenors in other proceedings to challenge the adequacy of ECCS regulations, and within a short time, led the AEC to convene hearings on the ECCS issue.

During most of the 1980s, a "beauty contest" mentality emerged from a perceived competition between RELAP5 and TRAC. The USNRC encouraged this perception believing that the competition would be healthy for the advancement of the Best-Estimate codes. They were right. Both teams worked intensely to show that their code was better. However, by the end of the decade, RELAP5 would emerge as the industry favorite primarily because of its relatively "user-friendly" features and execution speed.

While most of the 1980s was spent talking about the next generation of commercial nuclear power plants, the 1990s would be about "putting your money where your mouth is". All around the world work has been underway to design, certify, and/or construct the next generation of nuclear power plants. Reactor suppliers in North America, Japan and Europe have nine new nuclear reactor designs at advanced stages of planning and others at a research and development stage.

The direction of future nuclear safety research efforts is anybody's guess. Historically, there has been the trend that as knowledge and experience of nuclear reactor technology has increased, emphasis has shifted from reliance on containment structures to safety through the improved design of the reactor plant itself. This trend will likely continue. Ultimately, the future of any science is defined by the needs of the present. Probably the primary issue that needs to be addressed is the need to assure the public that nuclear power technology is safe and needed. The advanced light water reactor designs significantly reduced the probably of a severe accident and include inherent safety mechanisms which reduce the opportunity of human error. Such improvements will go a long way toward securing nuclear power as a technically safe electrical generation source. Unfortunately, addressing the issue of public concern is as much technical as it is political. Advances in nuclear safety alone will do little to change public opinion; however, neglecting nuclear safety research could result in a huge loss of public confidence in this industry.

Another factor that will affect the future of nuclear safety is economics. As governments deregulate the power industry, competition will force utilities to operate as efficient as possible. The goal that nuclear power plants compete economically with conventional (fossil fuel) plants will continue to create strong pressures to reduce capital costs, to increase reactor power levels, to lengthen core life, to achieve more efficient performance, and to bring reactors closer to metropolitan areas. The direction of many of these forces may appear counter to reactor safety. However, steady advances in the technology have made it possible to replace some of the very conservative safety measures, taken originally because of ignorance, by more realistic and, at the same time, more economic measures.

New Words and Expressions

criticality [krɪtɪˈkælɪti]	n.	危险程度
adjourn [əˈdʒɜːn]	v.	延期，休会，换另一个地方
fission [ˈfɪʃən]	n.	裂开，分裂，[原] 裂变
unleash [ˌʌnˈliːʃ]	v.	释放
graphite [ˈɡræfaɪt]	n.	石墨
terminology [ˌtɜːmɪˈnɒlədʒi]	n.	术语学
inundate [ˈɪnʌndeɪt]	v.	淹没
backlog [ˈbæklɒɡ]	n.	订货
coolant [ˈkuːlənt]	n.	冷冻剂，冷却液，散热剂
moratorium [ˌmɒrəˈtɔːriəm]	n.	延期偿付，延期偿付期间
intervenor [ˈɪntəˈviːnə(r)]	n.	干预者，介入者

科技英语结论的写作要点

科技论文的结论是在论文正文的最后，位于论文的讨论或结果与讨论部分的后面。论文的结论部分通常应包含以下内容：

① 陈述研究的主要认识或论点，包括重要发现或结果、发现或结果的重要内涵、对结果的说明或认识等。

② 总结性的陈述本研究结果可能的应用前景、研究的局限性以及需要进一步深入研究的方向。

应注意的是，书写结论时不应涉及正文中不曾指出的新事实，也不能在结论中简单地重复摘要、引言、结果或讨论章节中的句子；或叙述其他不重要甚至与本研究没有密切联系的内容。

一、语法要点

1) 在结论中叙述一些并不仅仅是自己特定的研究结果，而是普遍有效的结论时，通常使用一般现在时态。

- We cautiously conclude that these conditions seem to facilitate getting a better hold on accident prevention in companies.
- It verifies the importance of the safe behaviour model in being able to predict the safety

- performance of a mine and its workforce.
- The development of constitutive models is important for the understanding of soil behaviour as well as for the solution of engineering problems.

2）在结论中叙述一些重要的研究结果，结果的内容只是在本次特定研究的情形下才有效时，通常使用一般过去时态。
- It was indicated that the challenges of culture change needed to adapt to the current best practice in safety management.
- The criteria for the model were developed from the rules and regulations survey.
- A model was proposed which sought to integrate the main features of a safety management system.

3）在叙述进一步研究的方向或研究的题目，以及本次研究结果的应用或效益时，通常使用一般现在时态及may、should、could等情态动词。
- It is suggested that stress-state variables should not be mixed with volume-mass variables in deriving an equivalent "effective stress".
- Stress measurements should be made in the two walls of reverse faults in order to investigate any relationship to outburst occurrence.

二、结论常用句型

The results indicate/show/illustrate that...
结果表明/显示/说明……

Therefore, it was found that...
因此，（本文）发现……

The most important conclusion is that...
最重要的结论是……

The following conclusions can be derived from this work.
通过本次研究可得以下结论。

It was shown from the numerical simulation that...
数值模拟结果表明……

The results obtained from the investigation clearly demonstrate that...
本研究结果清楚表明……

It is possible to conclude that...
可得出的结论是……

Concluding the above discussion, we may state as follows.
通过以上讨论，我们可以得出以下结论。

From the above discusion it follows that...
从上述讨论可得……

The investigation of... leads us to the following conclusions.
对……研究使我们可以得出以下结论。

From the study the following recommendations should be made.
本研究提出以下建议。

Recommendation for further work are listed as follows.
进一步的研究如下。

There are several issues deserving further study.
有几点值得进一步研究。
Among further topic for future, we state...
进一步的研究课题，我们认为有……
The experiments mentioned in the paper confirmed that...
本试验证实……

Reading Material

Railway Safety Management

1. Introduction

Figures for transport in all its guises show that we are still in a period of major growth. Despite the attempts of some governments to discourage travel as one way of easing the congestion of roads, rail and air space, we can see annual growth of several percent in all modes. This is partly a result of the globalisation of business, generating much more transport to bring goods from the cheap labour economies to the high consumption societies of the developed world. Globalisation creates much more business travel also, as international companies try to keep control over their empires; travel which expands despite the possibilities of information technology to make the message travel without the person needing to. But also it is travel for pleasure which is driving the market; foreign holidays, a moneyed class of early retirees with time on their hands and a dispersion of families across countries. The challenge is to manage this growth with a commensurate increase in safety per kilometre travelled.

As a response to this railways are expanding rather than cutting lines. The high-speed train network is spanning Europe and forcing technical harmonisation as never before. Interoperability of trains on the underlying intercity networks is becoming increasingly vital. New forms of rail (or rail-like) transport are being introduced, from the Maglev (Transrapid, Swiss Metro) to the light rail hybrids with trams. Technologies such as tunnelling are being extended to new applications, such as the tunnels in the Netherlands to spare the above ground space or to reduce nuisance, and to new challenges, such as the Channel tunnel. GPS and other positioning techniques, together with better en route communication are changing traffic control techniques radically. New technology always brings with it new safety challenges.

Globalisation also means that national markets, such as for railways, are being opened up to foreign ownership, which brings with it new ideas of how to manage, including how to manage safety. Competition also is increasing, bringing other cold winds of change into the newly privatised railway companies. The response in many industries has been a major reorganisation of company boundaries, outsourcing much peripheral work, while core businesses seek to grow by acquiring their direct rivals, while at the same time splitting their own activities into (local) business units. Slogans such as "think global, act local" summarise this new concern for explicit competitive management. With new boundaries come new needs to define how safety can continue to be achieved in this rapidly changing world. I want to address these trends and challenges. First I want to look at how we define safety and then how we are changing the ways we manage it. The challenge to the railway industry is how to ride the crest of this wave of change in safety.

2. Approaches to Management

The last decade has ushered in the third age of safety. After the technical and human factors focus of the two previous ages, the dominate concern is now the company organisation and the role of its managers. How do they make sure that the hardware and people do what they should, when they should, to keep all risks under control? Making the role of management explicit has resulted from, but in turn also results in, a shift from reactive to proactive concern for safety. Management is all about planning and control. It looks forward and tries to anticipate and avoid problems. It is future-oriented and driven by clear goals and the need to survive and prosper. The textbooks of the management gurus are all about infusing this orientation for market success and survival, for quality and for economic success. By linking safety to this engine also, we try to take advantage of its drive. In this section this shift will be made explicit for a number of areas.

(1) From One-Off Fixes to Continuous Attention

One of the themes that we can read into the concern for technology and hardware safety in the first age of safety is the belief that we can design intrinsically safe processes. If we get it right at the beginning, we can then stop worrying about safety and get on with the important things in industry like operations and making profits. The early inspectors of factories, mines and railways were usually engineers, whose main interest was in the technology. They concentrated on the technical failures in accidents and on technological fixes for them. This focus can be seen in more modern times in the almost knee-jerk reaction to automate a process if the people working in it make a lot of errors and have accidents. I would not want to denigrate or undervalue the need to focus on hardware and system design and to adopt intrinsic safety and the elimination of hazards as a first option. However, almost two centuries of experience indicates that this is only a starting point. Ergonomics and cognitive psychology have shown that this approach has its limits and can add problems elsewhere (e.g. automated systems are more complex and unpredictable in emergencies, danger may be displaced from operators to maintenance staff). The history of automatic train control is a good case of this continuing battle to optimise the role of the driver and the technology for safety. Our experience is that the technology-fix does not stay fixed.

The second line of defence in many systems, if the human could not be eliminated, has been to try to turn the human into a robot by specifying rules and imposing them rigidly. The railway industry has been one of the main protagonists of this approach, alongside the nuclear, and to lesser extent, the chemical industries. Accidents were then analysed up to the point where it became clear that someone had broken a rule (at which point discipline was appropriate) or that there was no rule for this eventuality (in which case a new one was made). In this way rulebooks continually grew and never diminished. This rules-fix is also a hankering after certainty. Ultimately we get a rule for everything and safety is seen as something which requires no thinking any longer, but simply good training, a prodigious memory, a large safety manual or computer to refer to, and an iron discipline. Management does not need to do any more thinking or planning, because it is all fixed in the rule system.

What has killed the belief in the one-off fix as the whole answer has been the increasing pace of technological, market and organisational change. The system is never stable and so no fix will stay fixed. We constantly have to adjust and steer the system to adapt it to the changes. The emphasis therefore is on self-regulation and the ability to function as a learning organisation.

The difficulty we have for the future is to decide how much and what we should set down in rules, but above all how we can audit whether an organisation is managing its safety well, if it does not have a massively detailed rule book. We have to avoid the mistake of writing detailed management systems, which are simply replacing the rules for workers with equally inflexible rules for managers. Some audit systems make that mistake and impose detailed ways of managing, rather than emphasising a set of functions and tasks which must be carried out in whatever way the organisation (and accumulated good practice elsewhere) finds best.

(2) **Making Safety Explicit**

If managers are to manage safety consciously and proactively they must have performance indicators to do that. The indicators quoted in the first section of this paper are the old reactive ones of accidents. They are still necessary to act as a final check as to whether safety is improving or not, but we cannot steer by them. For process safety (passengers and goods) the figures are already too low to show reliable trends in many of the smaller operating enterprises. Soon they will only make sense as trends if the data for several countries are combined. This gives rise to the need for indicators related to earlier steps in the processes which finally lead to accidents if not blocked and recovered. We can look to near miss and incident data, hardware failure and maintenance data, behavioural monitoring, operational anomaly logging, inspection of the presence and working of preventive measures, audits and management reviews. The possibilities are legion. Railtrack is an example of a railway company which has gone far along this path in its safety plans and reports.

This development is to be praised and encouraged. Other rail companies should be following suit, both to give their own managers targets and to demonstrate to the public and government that they are serious about safety objectives and are achieving results in managing it. We have ample evidence to know that managers respond to management by objectives and alter their behaviour according to what is rewarded.

The industry needs to use its working parties to establish best practice and spread good ideas on viable performance indicators. Another motivation which has been driving the process of making safety explicit is the privatization and decentralisation of many railways. Instead of one monolithic company in which staff moved around from department to department and often stayed for their whole carriers, there is now a set of separate companies with less permeable boundaries. In the old system safety issues could permeate activities without being made explicit. Now they have to become the subject of contractual agreements between independent companies, to resolve liability issues. They are even becoming the subject of internal contracts between senior and middle managers (NS). They are also the basis of explicit safety cases in which companies bid for franchises or contracts, or which are formally assessed before a company receives a license to operate. Because these developments make safety explicit and controllable, perhaps we need to fear less the possibility that privatisation will increase the conflicts between the safety and economic performance goals of companies to the detriment of the former.

(3) **From Authoritarian to Participative**

Another concomitant of the move from the technical- or rules-fix to a continuous process of control and learning is that it demands much more participation. Rules can be imposed from above. They engender the culture that the experts know best and that the shop floor should be, or at least should play, dumb and simply obey the rules. Creativity and initiatives from the shop floor to improve are stifled in such a climate. Once rules become more conceptual and goal-oriented, rather than prescriptive, this cannot work. Nor can it work if the technology is changing or there are many abnormal situations to be coped with. It is hard to have an expert around all the time to make up new rules or modify old ones. The only other alternative is to delegate the rule making and changing to the work group. There needs to be a feedback loop to ensure that the written rulebook keeps up with the changes, so that new trainees are not taught old knowledge. However, we have radically altered the location of initiative and power from topdown to bottom-up. Studies of so-called High Reliability Organisations show that this is one of their characteristics. They have very explicit and central safety philosophies which emphasise the value of safety, but they do not have detailed rulebooks. The operating personnel constantly evolve, modify and pass on the safe working practices. Safety is a constant topic of conversation and incidents are informally reviewed to extract learning from them, making safety a live issue the whole time. There is also a high degree of cross-checking between individuals, creating an operational redundancy.

(4) Learning and Knowledge Management

If we put together all of the trends sketched, they add up to a revolution in the way safety is, or should be, managed in railway companies compared to a decade ago. The keywords are: systems thinking, explicit risk comparisons, integrated management systems across different types of harm, continuous concern for safety from management, participation from operational levels and finally the need to develop into a learning organisation in order to manage safety in a changing world. This adds up to a great demand for knowledge management in the rail enterprises and in the industry as a whole.

The need for knowledge management is accentuated by the breaking up of the national monopolies. The dedicated railway employee with a career wholly inside the industry will become less the rule. Already companies with no rail experience are bidding to run lines in countries such as the Netherlands. The first licence was issued to a company which had until recently run canal boats. There is also an increasing interchange at technical and higher management levels between the rail industry and other technologies. This provides some fresh winds of change and challenges to unquestioned old ways. It also requires that the knowledge about how to run safety needs to be made much more explicit, so that it can be learnt by such newcomers quickly, and not absorbed by an almost invisible process of osmosis over the years. Rail companies need to explore ways to accelerate the process of learning on the job, by developing explicit courses in safety management, philosophy and practice. Here too it may pay to open the door to collaboration with other industries, or to send staff on open courses run at universities or other training establishments.

New Words and Expressions

congestion [kən'dʒestʃən]   n. 拥塞
nuisance ['njuːsəns]   n. 损害
splitting ['splɪtɪŋ]   adj. 爆裂似的，极快的
intrinsically [ɪn'trɪnsɪkəli]   adv. 从本质上（讲）
knee-jerk ['niːdʒɜːk]   adj. 下意识的，自动反应的
denigrate ['denɪɡreɪt]   v. 毁誉
protagonist [prəʊ'tæɡənɪst]   n. 领导者，积极参加者

hankering [ˈhæŋkərɪŋ]	n.	渴望
prodigious [prəˈdɪdʒəs]	adj.	巨大的
permeable [ˈpɜːmɪəbl]	adj.	有浸透性的；能透过的
concomitant [dənˈkɒmɪtənt]	n.	伴随物
stifle [ˈstaɪfl]	v.	使窒息；抑制
accentuate [ækˈsentjʊeɪt]	v.	强调，着重强调

参 考 文 献

[1] Arto Kuusisto. Safety management systems: audit tools and reliability of auditing [D]. Tampere: Tampere University of Technology, 2000.
[2] Brian A J, John C B, Susan R M, et al. Emergency Resonders: safety management in disaster and terrorism response [R]. California: RAND Corporation, 2003.
[3] Roland. System safety and management [M]. New York: J. Wiley Co., 1990.
[4] Nelson & Associates. Core principles of safety engineering and the cardinal rules of hazard control [C]. http://hazardcontrol.com/coreprinciples.html, 1997.
[5] William S M, W Gary Allread. How to develop and manage an ergonomics process, institute for ergonomics [D]. Ohio: The Ohio State University Columbus, 2005.
[6] Hunszu Liu. Implementation of human error diagnosis (HED) System [J]. Journal of the Chinese Institute of Industrial Engineers, 2004, 21 (1): 82-91.
[7] Willie H, Dennis P. Occupational safety management and engineering [M]. 5th ed. London: Pearson Education Inc., 2001.
[8] Evelyn Ai Lin Teo, Florence Yean Yng Ling. Developing a model to measure the effectiveness of safety management systems of construction sites [J]. Building and Environment, 2006 (41): 1584-1592.
[9] Osama Abudayyeh, Tycho K F, Steven E B, et al. An investigation of management's commitment to construction safety [J]. International Journal of Project Management, 2006 (24): 164-174.
[10] N A Kartam, Flood I, Koushki P. Construction safety in Kuwait: issues, procedures, problems, and recommendations [J]. Safety Science, 2000 (36): 163-184.
[11] Brown. Construction safety & environmental management program [M]. Providence: The Office of Environmental Health & Safety, 2003.
[12] International Labour Office. Safety and health in construction [M]. Geneve: International Labour Office? 1992.
[13] Mohame S. Empirical investigation of construction safety management activities and performance in Australia [J]. Safety Science, 1999 (33): 129-142.
[14] Jose D P. Construction safety management (a systems approach) [M]. Dublin: Lulu, 2005.
[15] Dwyer T, Raftery A E. Industrial accidents are caused by the social relations of work—a sociological theory of industrial accidents [J]. Applied Ergonomics, 1991 (22): 17-178.
[16] Cook T M, Neumann D A. The effects of load placement on the activity of the low back muscles during load carrying by men and women [J]. Ergonomics, 1987 (30): 1413-1423.
[17] Crockford G W. Protective clothing and heat stress: introduction [J]. Annals of Occupational Hygiene 1999 (43): 287-288.
[18] Geller E S. The psychology of safety—How to improve behaviors and attitudes on the job [M]. Pennsylvania: Chilton Book Company, 1996.
[19] Nicola Healey. Analysis of RIDDOR machinery accidents in the UK printing and publishing industries 2003-2004 [R]. [S. L.]: Health and Safety Labor, 2006.

[20] Duzgun H S B. Analysis of roof fall hazards and risk assessment for Zonguldak coal basin underground mines [J]. International Journal of Coal Geology, 2005 (64): 104-115.

[21] Singh P K. Blast vibration damage to underground coal mines from adjacent open-pit blasting [J]. International Journal of Rock Mechanics & Mining Sciences, 2002 (39): 959-973.

[22] Szwedzicki T. Quality assurance in mine ground control management [J]. International Journal of Rock Mechanics & Mining Sciences, 2003 (40): 565-572.

[23] David Laurence. Safety rules and regulations on mine sites—the problem and a solution [J]. Journal of Safety Research, 2005 (36): 39-50.

[24] María B, Díaz Aguado, González Nicieza C. Control and prevention of gas outbursts in coal mines, Riosa—Olloniego coalfield, Spain [J]. International Journal of Coal Geology, 2007 (69): 253-266.

[25] Michael J S, Kenneth L C. Reducing the danger of explosions in sealed areas (gobs) in mines [N]. Milestones in Mining Safety and Health Technology. http://www.cdc.gov/niosh/mining/pubs/pdfs/tn489.pdf, 2001.

[26] Collins M, Davis B, Goode A. Steady state accommodation and VDU screen conditions [J]. Applied Ergonomics, 1994 (25): 334-338.

[27] Larsson T J, Rechnitzer G. Forklift trucks—analysis of severe and fatal occupational injuries—critical incidents and priorities for prevention [J]. Safety Science, 1994 (17): 275-289.

[28] Chao Elaine L, Henshaw John L. Job hazard analysis [R]. Washington DC: Occupational Safety and Health Administration, 2002.

[29] Ferrie J, Marmot M, Griffiths J, et al. Labor market changes and job insecurity: a challenge for social welfare and health promotion [R]. Denmark: Regional Office for Europe of the World Health Organization, 1999.

[30] Mayhew C. Getting the message across to small business about occupational violence and hold-up prevention: a pilot study [J]. Occupational Safety and Health, 2002, 18 (3): 311-325.

[31] Mayhew C. OHS challenges in Australian small businesses: old problem and emerging risks [J]. Safety Science Monitor, 2002 (6): 26-37.

[32] Mayhew C. Barriers to implementation of known OHS solutions in small business//National Occupational Health and Safety Commission and Division of Workplace Health & Safety [R]. Canberra: AGPS, 1997.

[33] Roger B. Strategy for meeting the occupational safety and health needs of small and medium size enterprises (SMES) a summary of ROSPA'S views [J]. Safety Science Monitor, 2003 (7): 1-13.

[34] Cromie S, Stephenson B, Monteith D. The management of family firms: an empirical investigation [J]. International Small Business Journal, 1995, 13 (4): 11-34.

[35] Waldemar K, William M. Fundamentals and assessment tools for occupational ergonomics [M]. New York: CRC Press, 2006.

[36] Reese C D. Accident/Incident prevention techniques [M]. New York: Taylor & Francis Inc., 2001.

[37] 韩其顺, 王学铭. 英汉科技英语翻译教程 [M]. 上海: 上海外语教育出版社, 1997.

[38] 俞炳丰. 科技英语论文实用写作指南 [M]. 西安: 西安交通大学出版社, 2003.